GROUP TECHNOLOGY
IN THE
ENGINEERING
INDUSTRY

GROUP TECHNOLOGY IN THE ENGINEERING INDUSTRY

A report on research, financed by the Science Research Council, by Birmingham, Bradford and Salford Universities, and by the London School of Business Studies

By JOHN L. BURBIDGE

MECHANICAL ENGINEERING PUBLICATIONS LTD

LONDON

First published 1979

ISBN 0 85298 402 2

Printed in England by Stephen Austin and Sons Ltd., Hertford

Bound by The Burlington Press (Cambridge) Ltd, Foxton Royston Hertfordshire

CONTENTS

ACKNOWLEDGEMENT

The permission of the Controller of HMSO to reproduce certain illustrations contained in this book is gratefully acknowledged.

LIST OF ILLUSTRATIONS

In some instances the illustrations spread over more than one page. Where this is the case the page number given refers to the page on which the illustration begins.

CHAPTER 1

SETTING THE
MANUFACTURING SCENE

1.1. INTRODUCTION

Britain, perhaps more than any other country, depends on international trade in manufactures. Manufactured goods constitute a large percentage of UK exports, and engineering products provide a high proportion of this total. Capability in engineering product design and in the management of production is essential if Britain is to survive.

The contribution to our economy of the engineering industry is, however, much less than it must be if Britain is to maintain its standard of living. The industry is at present running at levels of productivity which are lower than those maintained in other countries with which it competes.

This book reports the results of research, financed by the Science Research Council, into Group Technology in the engineering industry. It is mainly concerned with parts manufacture for engineering products. In other words it looks principally at the problems of operating machine shops which make engineering components. The book shows that in Group Technology there is a new approach to work organisation which can help to solve many of the industries present problems.

This first chapter is based mainly on a study made at the London Graduate School of Business supported by sociological studies made at Birmingham University. It looks at the problems facing the engineering industry in Britain and attempts to evaluate its performance in comparison with some of its main competitors.

These studies see the main problems facing the engineering industry as originating, not in the environment in which companies work, but rather inside the factories, particularly in the field of production management. They see the main cause of the problems in production management as arising from the complexity induced, firstly by over-specialisation in manufacturing sub-processes and in minor manage-

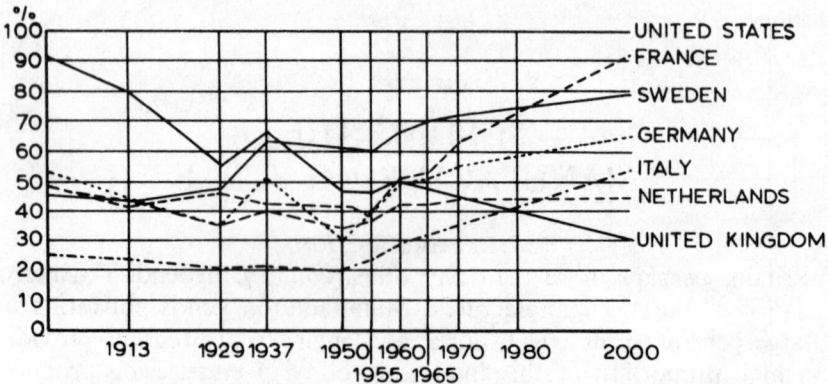

(a) *Historical trends in GNP, per capita, 1899-1980 (in percentage US GNP, per capita, computed from World Bank data. At factor cost and 1969 US $. Graph only shows historical trends)*

Fig. 1.1 Gross National Product (GNP)

(*Reproduced from* The United Kingdom in 1980: The Hudson Report (*Associated Business Programmes Ltd., 17 Buckingham Gate, London, SW1*)

ment functions and techniques, and secondly by excessive growth in the size of production units. They note as a contributory cause of growing importance, changes in the social system which are tending to make traditional systems of factory organisation obsolete.

A need is seen for new systems of organisation for production and new methods of planning work, which will both simplify production and make work more satisfying for the people in industry. Group Technology is indicated as a new approach to production organisation which can help to fill this need.

1.2. THE BRITISH ENGINEERING INDUSTRY
The alarming facts of Britain's decline in manufacturing performance are spelt out clearly in many recent publications. Here it is only necessary to outline the main facts.

1.2.(a) Relative decline in gross national product per head
Figure 1.1 (a) shows the UK's relative decline in terms of gross national product per head of population during the current century, compared with other European countries, using United States performance as the standard. Analysing this growth pattern for the UK by industry over recent years, Fig. 1.1 (b) shows the recent pattern

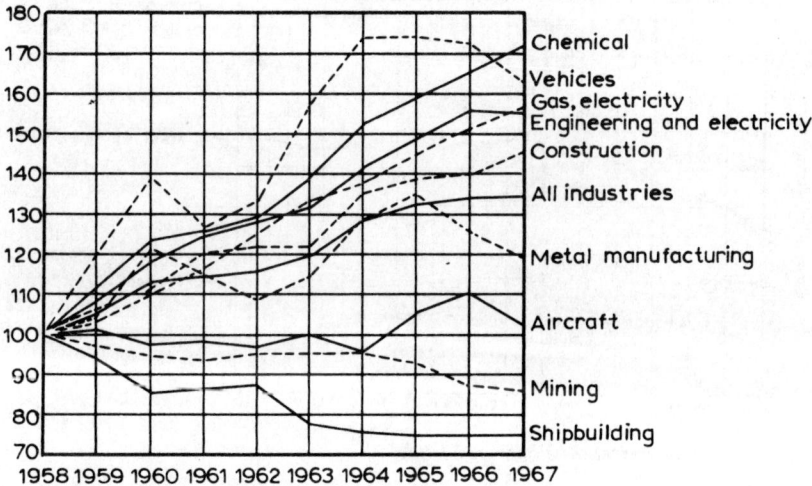

(b) *UK output trend by industrial sector*
(*Reproduced from* The Role of Mechanical Engineering (*HMSO*))

Fig. 1.1 Gross National Product (GNP)

of expansion and contraction in the major manufacturing sectors. The diagram displays a marked decline in ship building which was surely unnecessary considering the level of technology, historical position and trading interests of the UK.

1.2.(b) Poor achievement in exporting
The importance of the export problem is dramatically illustrated by the steadily worsening balance of payments in the UK now running into thousands of millions of pounds.

Britain's achievement in exporting is not impressive. Figure 1.2 (a) shows changes in the percentage of world trade over the two decades ending in 1971, of the major industrial countries, together with the percentage increase in total exports by country. Analysing these figures further, Fig. 1.2 (b) shows the UK export levels in 1967 for the major manufacturing sectors. From these figures, engineering in its different categories can be picked out as the most important exporting sector.

It is not enough, however, simply to export in volume. The value of the exported goods is also important. Figure 1.2 (c) shows the

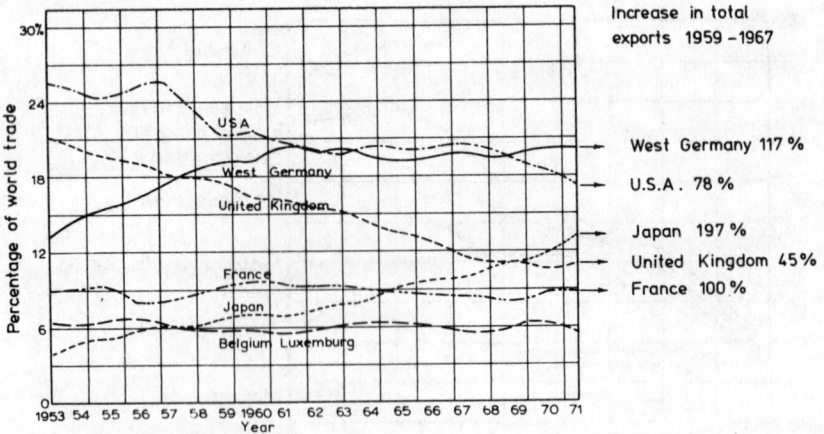

(a) *Changes in percentage of world trade*

UK exports 1967: Total £4,750 million

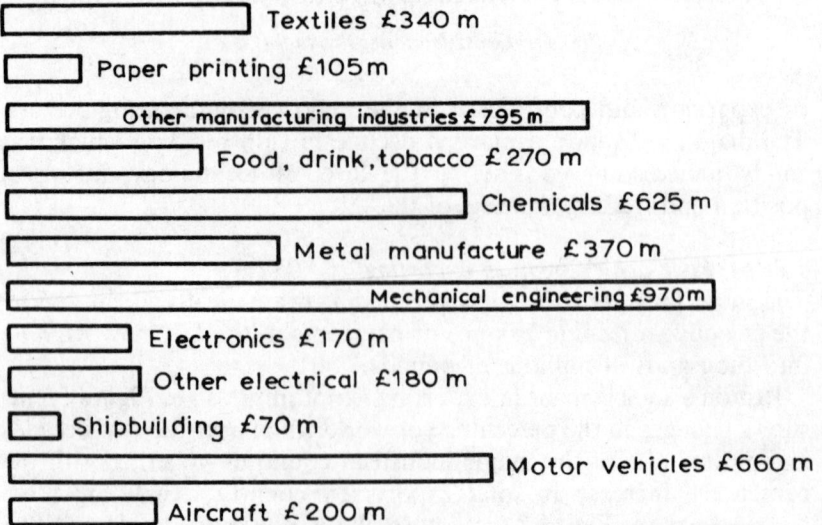

Textiles £340 m

Paper printing £105 m

Other manufacturing industries £795 m

Food, drink, tobacco £270 m

Chemicals £625 m

Metal manufacture £370 m

Mechanical engineering £970 m

Electronics £170 m

Other electrical £180 m

Shipbuilding £70 m

Motor vehicles £660 m

Aircraft £200 m

(b) *UK exports by manufacturing sector*
(Source: *Department of Trade and Industry*)

Fig. 1.2 UK exports

(*Reproduced from* The Role of Mechanical Engineering (*HMSO*)

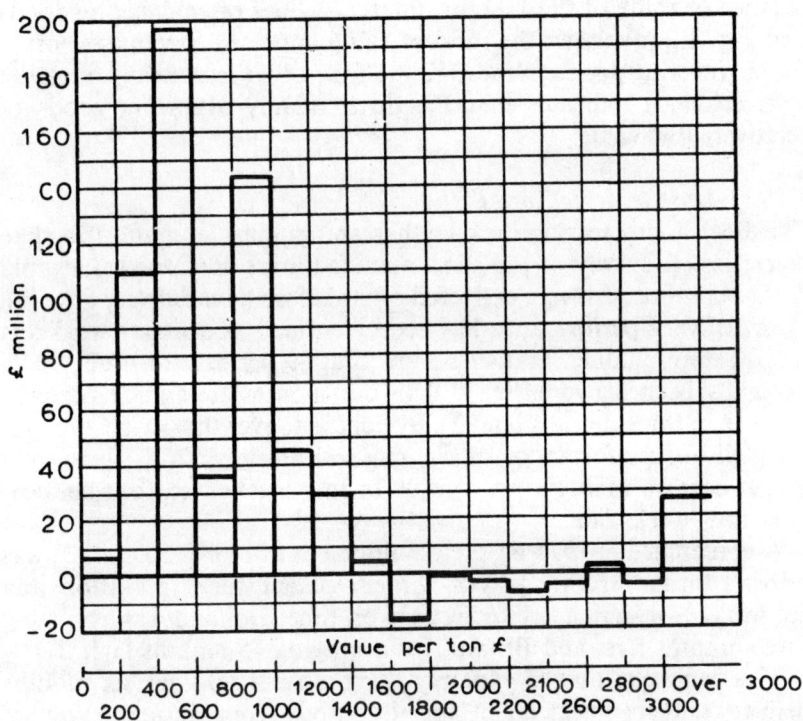

(c) *UK Balance of Trade by value of engineering goods*

Value per tonne:	1. Exports	2. Imports	3. Ratio 2/1
Sweden	2709	2269	0·84
W. Germany	2589	2556	0·99
France	2333	2559	1·10
Italy	2339	2791	1·19
UK	2040	3808	1·87

(d) *Ratio: value of engineering goods imported to exported, 1969*

Fig. 1.2 UK exports

6 *Group Technology in the Engineering Industry*

balance of trade of the UK for different values of engineering goods and Fig. 1.2 (d) shows the ratio of value imported to value exported for engineering goods in the UK and four other European countries. This evidence indicates that Britain is mainly exporting goods of relatively low value.

1.2.(c) Unreliable deliveries

The British engineering industry has an apalling reputation for late deliveries. A survey of job shop manufacturers confirmed that only 3 per cent of orders were delivered on or before their delivery date (1). Poor delivery performance has been identified as a major weakness in UK ship building. Statistics show that 39 per cent of orders have normally been delivered more than one month late and 21 per cent more than three months late (2). By contrast, over the period 1967–71 the delivery performance of six major European ship yards was: 90 per cent on time, six per cent up to four weeks late, four per cent over four weeks late.

A conference in 1975 on the development of North Sea oil (3) was informed that in the delivery of equipment for exploration and recovery, Norwegian deliveries were on time, Holland was averaging three months late, and Britain was averaging 15 months late.

The comment that if you buy from Britain you get six months delivery and six weeks credit and if you buy from Germany you get six weeks delivery and six months credit is at least indicative of our present reputation.

1.2.(d) Excessive stocks

Comparative studies between USA and UK companies have shown that UK companies require twice the investment in stock per unit of output, compared with their American equivalents.

Anthony Vice found that the British General Electric Company (GEC) compared unfavourably with American companies in their stocks position, although it compared well in other factors (4). 'Even with an efficient company like GEC, the ability to sustain the American General Electric Company's stocks/sales ratio would enable it to cut inventories by some £25 million. For UK industry as a whole this would imply a possible inventory saving of perhaps £1750 million, a massive once for all gain for the balance of payments and bank financing.'

	Investment %	Output %
USA	7·5	4·5
Belgium	9·8	5·9
France	9·6	7·0
UK	4·4	3·5
Japan	16·0	13·0
Canada	5·8	5·5
W. Germany	5·0	6·5
Sweden	4·1	6·3

(a) *Growth rates of investment and output*
(*Manufacturing industry 1959-69, percentage at constant prices*)

	Under 5 years %	Under 10 years %
UK	19	41
Italy	25	50
Germany	35	65
Japan	32	63

(b) *Age structure of machine tools in use*
(Source: Metalworking Production. *Third survey of M/C tools and production in Britain*)

Fig. 1.3 Investment

1.3. ECONOMIC THEORIES AND EXCUSES
Economists and others have put forward a number of theories to explain the present deficiencies of British industry. Of these the following are the most widely promulgated.

1.3.(a) Low productivity is due to low investment
Low investment in new machinery and equipment is often put forward as a main cause of low productivity in the British engineering industry.

That investment is relatively low in the UK is evident from an examination of Fig. 1.3 (a), showing the rate of growth in investment and output in a range of developed countries. That investment is low is also supported by the comparative age structure of machine tools in use, which is given in Fig. 1.3 (b).

	Pre-Tax		Post-Tax	
	UK %	W. Germany %	UK %	W. Germany %
1961	13·3	13·2	9·5	5·2
1962	11·9	12·6	8·7	5·1
1963	12·6	11·8	9·2	4·8
1964	14·2	12·5	10·1	4·9
1965	13·4	11·3	9·1	4·7
1966	11·8	10·3	7·6	4·6
1967	11·9	9·8	7·3	4·9
1968	13·7	11·4	8·3	5·0
1969	13·1	11·8	7·8	5·4
Av. annual change 1961-1969	0·0	−0·2	−0·2	0·0

(c) Rates of return on capital. UK and W. Germany
(Manufacturing industry, historic cost)

Fig. 1.3 Investment

The cause of this low investment is obscure. It is normally thought to be caused by low rates of return on capital in British industry, but this view is not supported by the facts. Comparing the UK with West Germany, which had a high rate of investment in the decade to 1969, it can be seen from Fig. 1.3 (c) that in every year but one, the UK had a higher pre-tax rate of return on capital, and that the rate of return after tax, was higher in every year.

A curious fact is that the sources of finance which make investment possible, tend to favour the larger companies, which in general tend to have the worst labour problems and to achieve a relatively low return on capital investment compared with smaller companies. In a study of the financing of industry, Spiegelberg (5) perhaps identified the cause of this favouritism in the allocation of funds, when he stated: 'The main necessary condition for plausibility in the City, hangs around the social ease with which the leaders of a company seeking financial assistance can present themselves'. This is the sort of skill which may be thin on the ground in a new, small, aggressive enterprise, with good control of its labour force.

Investment in new plant does generally increase labour productivity.

The profitability of such investments depends, however, on the organisation of the company concerned. The present research has shown that a very small investment in organisational change can induce major increases in productivity and profitability with the existing plant and also increase the profitability of later investments in new plant. It seems reasonable to give priority to an investment which can give a high initial rate of return, coupled with an increase in the profitability of future investments.

1.3.(b) Other nations with their own natural resources are catching up industrially

Other commentators argue that Britain's weakening export performance is inevitable, due to other nations which have their own natural resources catching up in industrialisation. This argument would have been more convincing if Japan and Germany, with no greater natural resources than the UK, had not been able to achieve major industrial successes since the last world war.

1.3.(c) Too many products

Another theory is that British industry is uncompetitive because too many plants manufacture too wide a range of products, leading to short runs and less sophisticated machinery. It has been said that many industries need rationalising. However, the rationalisation phase of the late 1960's supported by the Industrial Re-organisation Corporation, did not result in conspicuous successes. Consider for example the British automotive, aero-engine and computer industries.

The present research, in fact, seems to indicate that the reverse is true, and that many of our present troubles spring from excessive growth in the size of companies coupled with excessive centralisation of control. Increases in company size may increase financial and marketing strength, but they don't appear to increase production efficiency.

1.3.(d) The shift into non-manufacturing employment

Some economists see the shift of workers out of manufacturing as a major factor in explaining Britain's exporting weaknesses. The argument is that services, mainly public sector activities, have been allowed to grow too fast and have absorbed too much of the potential and actual manufacturing capability. A proposed solution is to cut or restrain the public sector.

Excessive spending in the public sector has certainly diverted investment capital away from industry and has depressed the economy. As far as the effect of people moving out of industry on the efficiency of existing companies is concerned, however, the argument would be more convincing if there were a large unfilled demand by industry for new labour, and if there were not over a million unemployed. A further paradox exists in the widely held belief that at present levels of output, British Industry is still heavily over-manned.

It is interesting to note the relative movements of labour in Germany and the UK. As Germany has moved towards a higher investment in manufacturing than the UK, it might be expected that more labour would have moved out of their manufacturing sector than here. In fact by the end of 1962–72 decade, 30 per cent more of the UK labour force worked outside manufacturing than at the start of the decade. This must be compared with the movement of only eight per cent in Germany.

1.3.(e) Industrial unrest
Another theory sees our 'failure in industrial relations' and lost mandays due to strikes, as a major cause of lost productivity. In fact the number of days lost through strikes is not as high as might be expected from the emphasis laid on the subject by the Press.

Figure 1.4 (a) shows that the number of days lost through strikes has grown steadily in the UK since the war, but comparing the number of working days lost per 1000 employers in other manufacturing countries, Fig. 1.4 (b) demonstrates that the British figures are far from the worst in the world. Figure 1.4 (c) shows that a high proportion of our labour problems are concentrated in the larger nationalised industries.

It must also be remembered that the number of days lost through strikes is a small proportion of the days lost through unemployment, injury and sickness. In 1968 for example, although five million man days were lost through strikes, 146 million days were lost through unemployment and 324 million were lost due to sickness and injury.

Strikes are certainly a symptom indicating that something is wrong in British industry. The fact that more days are lost per employee in

	Number of strikes	Workers involved ('000)	Striker-days ('000)
1900-10	529	240	4 576
1911-13	1074	1034	20 908
1914-18	844	632	5 292
1919-21	1241	2108	49 053
1922-25	629	503	11 968
1926	323	2734	162 233
1927-32	379	344	4 740
1933-39	735	295	1 694
1940-44	1491	499	1 816
1945-54	1791	545	2 073
1955-64	2521	1116	3 889
1965	2351	868	2 932
1966	1937	530	2 395
1967	2116	731	2 783
1968	2378	2255	4 719
1969	3116	1654	6 925
1970	3906	1793	10 908
1971	2223	1173	13 558

(a) *UK strike statistics. Annual averages 1900-71*
(Source: Department of Employment Gazette)

Country	Days lost
Sweden	20
W. Germany	23
Netherlands	25
Belgium	226
N. Zealand	235
Japan	236
France	306
UK	321
Denmark	414
Australia	464
India	813
Italy	1003
USA	1121
Canada	1207

(b) *Working days lost per 1000 workers due to disputes. Yearly average figures, 1960-70*

Fig. 1.4 Strikes

Group Technology in the Engineering Industry

	Annual average number of man-days lost through disputes				Average number of man-days lost yearly in disputes per thousand insured persons			
	1930/38	1947/55	1956/64	1965/68	1930/38	1947/55	1956/64	1965/68
Docks	32 800	344 400	169 100	238 700	285	3134	1091	2325
Shipbuilding & shiprepairing	54 200	194 100	514 600	176 200	328	890	2349	869
Coalmining	1 002 600	616 100	444 000	173 700	1034	778	627	298
Engineering & vehicles	88 000	441 700	1 290 300	1 206 700	80	162	411	402
Construction	71 600	87 900	172 000	203 000	60	69	110	130
Textiles	1 504 000	21 000	27 100	33 700	1311	22	30	41
Food, drink & tobacco	5 400	12 000	22 400	29 700	10	15	27	35

(c) *Man days lost in seven major industries*
    (*Source: Devlin Committee report on the Port Industry 1963. Cmnd 2734*)
                        Fig. 1.4 Strikes

very large enterprises with highly centralised management, than in
smaller companies, supports the thesis that over specialisation and
excessive growth in the size of production units are major causes of
our present difficulties.

1.3.(f) Educational and social causes
Other commentators suggest that the long decline of the British
industrial position is attributable to educational and social factors.
They stress the low status and educational levels in the engineering
and management professions.

A Department of Industry study (6), comparing the origins and
attitudes of European managers in the 1970's, found that German
manufacturing management has a high social standing, that their
managers are well educated and well motivated, and that industrial
careers are more prestigious than careers in the cultural and academic
professions. They found that in France, engineers constituted the
most heavily represented group at the top of industry, and that nine
out of every ten of a sample of chief executives had had a university
level training. Similarly in Sweden they found that the typical entrant
into manufacturing management was an engineering graduate who
began his career in production.

By contrast British managers were found to be 'comparatively
poorly educated' and technical men were poorly represented at the
higher levels of management. The study report stated that Britain's
higher educational institutions, unlike some continental and North

American ones, have failed to provide industry with ambitious and able generalists with technical and commercial qualifications. The destination of technical talent in Britain has been to pure science and research, rather than to technology.

This theory depends in part on the theory that good academics make good managers, a theory which is perhaps not completely proven. British production managers generally have had more practical shop floor experience than their continental counterparts. It can be argued that the knowledge and skills gained by experience can be as valuable as academic knowledge and that the main reason why British production managers are not effective, is that they are not being properly used.

It can be argued too, that British universities have very limited capacity for training in production management—our area of greatest weakness—and that the industrial systems in some countries in which promotion is limited to those with the 'right' degrees, are socially divisive and may be unstable in the long run.

1.4. PARALYSIS OF THE OPERATING SYSTEM

Economists have looked for the causes of Britain's industrial difficulties in the environment in which companies work. Their theories have indicated areas of weakness where improvements might be made, but have not been convincing as discovering the major causes of the difficulties. The present research indicates that a more probable cause can be found inside the companies in industry. It is argued that the main cause of our difficulties in the engineering industry is paralysis of the production systems due to excessive process specialisation and excessive growth in the size of production units. These factors have so increased the complexity of production systems, that many engineering companies are now out of control.

1.4.(a) Out of control

The management of a production system might fairly claim that they had control of the system if, say, 90 per cent of their plans were achieved in practice.

This would mean in an engineering company, for example, that if the production programme showed that 100 of product *A* were to be completed in period *X*, at least 90 would be completed. This in turn would require that at least 90 sets of parts must be available at the beginning of the assembly period. This again requires that materials must be available in time so that the parts can be machined before they are needed.

As written, this does not appear to be a very ambitious objective. The research has failed, however, to find any engineering company making assembled products which could claim that they consistently achieve this level of performance.

Companies have been found which on average only deliver 30 per cent of their products by promised delivery date; complete only 12 per cent of the shop orders issued to their machine shops by due-dates, and in which 75 per cent of their material deliveries from suppliers arrive late. These production systems are obviously out of control. What is more, there is evidence that some of these results are not special to a few bad companies, but are not far below the average for the industry as a whole. The research has shown that many companies in the engineering industry do not measure their performance systematically in these areas, and that some do not know that their production is out of control.

It is part of the thesis of this study, that these results are not due to deficiencies in the production managers who attempt to manage production, but are due to the systems themselves. Due to an excessive belief in process specialisation and 'bigness', production systems have been created which no one could hope to run efficiently.

1.5. SPECIALISATION

The great majority of the engineering companies in the world are organised on the basis of process specialisation. In other words the men together with their machines and equipment are divided into organisational units each of which specialises in a particular manufacturing process, or sub-process.

1.5.(a) Effect of process specialisation on material flow

In workshops which make components, this philosophy can be illustrated by the case of the typical engineering machine shop. In this case the shop is generally divided into different sections for lathes, for milling machines, for drilling machines, and so on. Each group of workers specialises in one process.

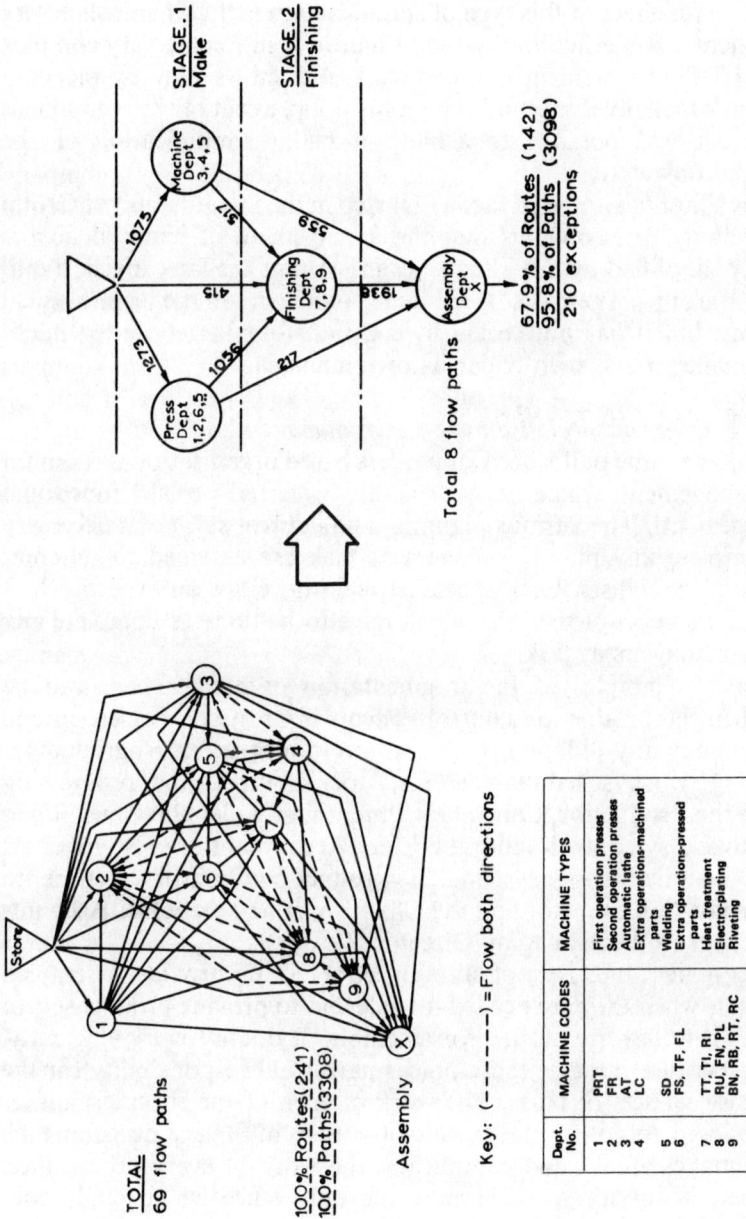

STAGE. 1
Make

STAGE. 2
Finishing

Machine
Dept
3,4,5

Finishing
Dept
7,8,9

Press
Dept
1,2,6,5

Assembly
Dept
X

1075

517

559

415

1273

1056

217

1938

Total 8 flow paths

67·9% of Routes (142)
95·8% of Paths (3098)
210 exceptions

TOTAL
69 flow paths

100% Routes(241)
100% Paths(3308)

Assembly

Store

Key: (– – – – –) = Flow both directions

Dept. No.	MACHINE CODES	MACHINE TYPES
1	PRT	First operation presses
2	FR	Second operation presses
3	AT	Automatic lathe
4	LC	Extra operations-machined parts
5	SD	Welding
6	FS, TF, FL	Extra operations-pressed
7	TT, TI, RI	Heat treatment
8	RU, FN, PL	Electro-plating
9	BN, RB, RT, RC	Riveting

Fig. 1.5 Simplification of material flow

The main effect of this type of specialisation is that it creates a very complicated material flow system. Figure 1.5 shows the very complex material flow system in one company and shows how by reorganisation, which involved modifying only 4·2 per cent of the component routes, it was possible to achieve a major simplification of the material flow system.

This simplification is in fact a first step in the introduction of Group Technology. It is obvious that the management of material flow is greatly simplified by this change, even without the later division into smaller groups. There is still the same complexity in the technological systems, but it has now been divided into four largely independent and smaller parts, each under its own manager.

1.5.(b) Effect of specialisation on staff managers

Exactly the same philosophy of process based organisation is followed in management where it is generally referred to as 'functional management'. The various planning and control sub-tasks necessary to complete any major management task are assigned to different teams of specialists. Each specialist team develops an expertise in a limited selection of the jobs which have to be done to complete any major management task.

Typical examples of the fragmentation of management can be found in the production control systems in most companies. In one typical company different 'offices' are responsible for programming, ordering, work scheduling, stores, stock control, and progressing under the Production Controller. Purchasing under the Chief Buyer and inventory control under the Chief Accountant are also involved in the material supply system. The production Controller reports to the Production Director and the Chief Buyer and the Chief Accountant report to the Managing Director.

The main objectives of all this effort is to provide assembled products when they are needed for sale and to provide product sets of parts when they are needed for assembly. It doesn't work efficiently, partly because none of these offices have total responsibility for the provision of sets of parts. The work of each of the specialist offices is supposed to contribute to this objective, but every decision each office makes affects and complicates the work of every other office. Further, it doesn't work because the only office which really considers the major objective of providing sets of parts, is the progressing office. Unfortunately their 'progress chasing' efforts make it impossible for any other office to achieve its limited objectives. Finally it doesn't work because no one lower than the Managing Director has the necessary authority over all the offices concerned with the material supply and processing activities, to ensure that sets or subsets of made and bought parts are available when needed.

This question of production control will be examined in greater detail in Chapter 8. It can be stated here, however, that production control is basically one of the simpler parts of management. It is only made complicated by the practice of basing its organisation on minor data-processing processes and techniques, rather than on the product. It is possible and usually more effective to have different offices each totally responsible for the provision of a different class of made or bought item.

Exactly the same principle of process specialisation is found in other management functions. Many production planning departments, for example, have separate sections for routing, time study, method study, tool design, and standards. One famous machine tool factory, has changed the organisation of its production planning department into multi-disciplinary planning groups, each totally responsible for methods and tooling in one or two production groups.

1.5.(c) Effect of specialisation on line management

Some degree of staff specialisation will always be necessary. In many companies today, however, specialist centralised planning and control has been extended to cover the minute details of how, where and when each operation should be done.

The impact of this type of specialisation has been to remove discretion in the management of operations from the line managers. Little flexibility is left for the line manager to exercise discretion on a daily basis. The interest, and imaginative challenge of running the system is taken away by the policies, rules, procedures and plans of the specialists. The main effect is immense frustration in the line staff because they are not allowed to make the appropriate flexible decisions in relation to current operating necessities.

	1935 %	1958 %	1963 %	1968 %
Shares of largest firms in employment				
50 firms	14·9	24·7	27·9	32·4
100 firms	24·0	32·3	37·4	42·0
200 firms	n.a.	41·0	47·9	52·5
Shares of largest firms in employment				
50 firms	15·0	21·2	24·3	29·4
100 firms	22·0	27·7	32·6	37·8
200 firms	28·0	35·5	42·0	47·1

(a) *Share of large firms in UK output and employment*
(Source: *M C Sawyer*, op cit, *and calculations from* Census of Production 1968, *ratio of 100 firms for output for 1968. S J Prais, 'The share of the largest 100 manufacturing firms in Britain 1960-70,—unpublished paper, NIESR, 1970*)

Fig. 1.6 Growth in size of companies

1.6. GROWTH IN THE SIZE OF PRODUCTION UNITS
The second factor contributing to the paralysis of the operating system has been the general increase in the size of companies, coupled with a trend towards centralisation of control.

1.6.(a) The growth of UK companies
It is seldom appreciated how much UK industry has become concentrated in the hands of fewer and fewer companies. Figure 1.6 (a) shows how the share of output and employment of the 200 largest British companies has grown over the last few decades. Figure 1.6 (b) shows how capital outlays for corporation acquisition have increased even more in the UK than in the USA.

The Bolton report **(8)** found that the share of manufacturing employment in small enterprises was 20 per cent in the UK compared with 30 per cent in the USA. A small enterprise was designated as one which employed less than 200 people. It also found that, in comparison with many other countries, Britain had the smallest proportion of total manufacturing output coming from small firms and that the contribution made to manufacturing by small firms in Australia, Switzerland and Norway was running at twice the level of that in Britain.

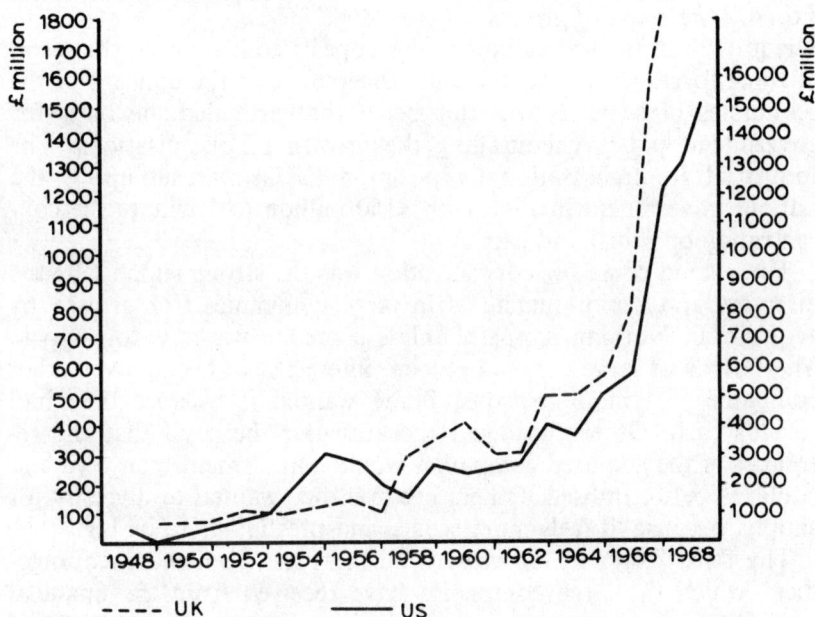

---- UK ———— US

(b) *Capital outlays on corporate acquisitions in USA and UK, 1948-68*

Fig. 1.6 Growth in size of companies

As companies have grown in size, so their formalised procedures have become centralised and 'rationalised' across their different operational sites. Thus decisions on such factors as product range, research and development, process choice, supplies sources, market outlets, the use of computers, employment, working conditions, and investment in new plant, are determined not by the operational site manager but by the staff at head office, wherever that may be. The overall control of the production system has been removed from the site managers and the factories have become centrally controlled head office departments as a consequence of centrally determined company policy. It is not a coincidence that Britain's few efficient large companies, such as GEC and BOC for example, are strongly committed to a division into small production units and to devolution.

1.6.(b) The cause of growth

The growth in the size of companies appears to have been the result
of three forces. First was the wide acceptance of the concept of the
economies of scale. It was this belief that provided the basis for
government policy encouraging the growth of organisations. The
Industrial Re-organisation Corporation (IRC) was set up by the
Labour government in 1966 with £150 million to facilitate the concentration of British industry.

The second cause for concentration was the strong support by the
directors and senior managers in large companies, for growth by
acquisition. Nothing happens unless someone wants it to happen.
The boards of large companies were interested in taking over other
companies for many reasons. Some wanted it because they had
surplus cash. Others wanted it because they believed that the resources of the acquired companies would complement their own and
could be better utilised. Others because they wanted to diversify, or
simply, because 'it feels more secure and prestigious to be big.'

The third factor in the growth business has been the encourage-
ment which the large companies have received from the financial
institutions. Growth in companies by acquisition requires financial
support and the lender of the funds must approve the objectives for
which his money is going to be used. As Speigelberg explains (5),
'Merchant Banks played an important role in selling the dubious
concept of bigness and the need for concentration to their industrial
clients. In doing so they were not only over-selling their own competence, they were succouring the megalomaniac ambitions of many
British industrialists.'

1.6.(c) Comparison of large and small companies

The main beneficiaries of growth should have been the large companies themselves. It is doubtful if most of them have achieved any
real tangible benefits. Reddaway analysed a number of takeovers (7)
and found that profitability generally decreased after a takeover.
Several of the companies which had been encouraged, by government,
to expand, had to be rescued later by further injections of government
finance.

Again, the Bolton report (8) found that the ratio of net profit to
net assets was 30 per cent higher for small companies than for large

organisations. One of the largest companies in the UK with many divisions, expects 60 per cent efficiency from its plant with 400 people and 80 per cent from plants with 200, and attributes the difference simply to size. Perhaps even more important as regards the UK, a NEDO study (9) showed that smaller companies had contributed more to growth in exports per pound of sales, than the large companies.

Doubts are often expressed about our large nationalised industries. The public may express doubts about the idealogy of nationalisation, but what they [the nationalised industries] are really suffering from is the inertia of large organisations; ONCL, the nationalised freight carrier, scored a success within a year of being decentralised into small regional units, by turning a loss of £2 million into a break-even position.

It seems that the main outcome of industrial and commercial concentration has been a reduction in competition rather than any improvement in productivity or in customer service in terms of design range, quality, price or delivery.

1.6.(d) The effect of growth on production management
One of the most important effects of the increase in size has been to complicate production management to a stage where in many companies it is now out of control.

With a few notable exceptions most large companies have built up their headquarters units, have centralised most of the planning and have attempted to standardise the policy followed in all their different plants. Policies, plans and procedures are formulated by specialists in the head office. Their decisions come together at the level of the floor of the factory. These policies and plans produced by different specialists represent a large number of sub-optimum solutions to the needs of production. Not unnaturally the total production system which evolves by this method is generally chaotic in its complexity.

The factory manager on the spot who might be able with his local knowledge to solve the problems, is unable to do so because he is largely constrained, by policies, budgets and imposed systems, from taking useful action. For factory managers the great loss has been in their independent responsibility and control. All the procedures, decisions and policies from remotely situated experts, act as constraints on their will and energy.

In addition to the increase in the number of different factories controlled by the larger companies, there has also been a trend towards consolidation into larger production departments inside factories. One result of this increase in the size of production units has been to increase the complexity of the material flow systems in factories. Attempts to overcome the problems of this complexity by centralised systems planning and computerisation have failed to solve the problems and have led to a large increase in the amount of paper work to be dealt with in the factories. Another result has been a deterioration in labour-management relations.

Most of our problems in the engineering industry today are in the production field. Production management is at the heart of these problems. Our main problems are not in marketing, or in accounting, or in personnel management: they are in the day to day control of the labour force and of the material supply and processing activities in our factories.

What is needed today is for production systems to be designed as simple systems, composed of relatively small, largely independent units at each level of management, to which the authority and responsibility for production performance can be clearly allocated. If this is done and simple objectives are established covering major requirements, the production managers and the men on the floor of the shop are more competent to make the system work and to overcome the random mischances of life in a factory, than will ever be possible with any combination of detailed formal systems designed by independent staff specialists. This approach to the organisation of work is called Group Technology.

1.7. THE POTENTIAL FOR CHANGE

There is considerable evidence that the British economy not only has plenty of reserves of physical capacity in terms of labour, machinery and management but also, in Group Technology, has methods available which can help in harnessing that capacity.

The evidence that there are reserves of capacity was never better displayed than during the government imposed three day week in the early part of 1974 when many manufacturing companies were

Industry	Activity						Productive individual firm %	
	Operating %	*Attending* %	*Handling* %	*Servicing* %	*Waiting—Management responsible* %	*Waiting—Operative responsible* %	*Lowest*	*Highest*
Iron and Steel	43·0	8·0	24·0	4·5	10·5	10·0	23·0	66·0
Non Ferrous	45·0	18·0	18·5	5·0	6·0	9·5	29·0	71·0
Motor Vehicle Components	55·0	10·0	15·5	4·5	4·5	10·5	42·0	60·0
Electrical	50·0	6·0	23·0	4·5	5·0	11·5	36·0	71·0
West Midlands Sample	48·0	11·5	19·5	5·0	5·5	10·5	23·0	71·0

(a) Utilisation of labour

Industry	Activity						Productive individual firm %	
	Productive %	*Setting* %	*Tool adjustment* %	*Maintenance* %	*Idle—Management responsible* %	*Idle—Operative responsible* %	*Lowest*	*Highest*
Iron and Steel	42·0	7·5	3·5	3·5	36·0	7·5	28·0	62·0
Non Ferrous	45·0	2·5	1·0	3·0	44·0	3·5	17·0	66·0
Motor Vehicle Components	45·0	3·0	2·5	1·5	41·0	7·0	32·0	65·0
Electrical	30·0	4·0	1·5	1·0	59·0	4·5	24·0	42·0
West Midlands Sample	41·0	4·5	2·0	2·0	45·0	5·5	17·0	66·0

(b) Utilisation of machines

Fig. 1.7 Analysis of potential added values
(Source: Norman and Bahiri (10)

Industry Group	Current (1968) Added value £	Potential Added value £ Labour	Machines	Increase % Labour	Machines
Iron and Steel	2146	3249	3168	53	48
Non-Ferrous	2058	3247	3018	57	47
Motor Vehicle Components	2106	2718	3042	30	44
Electrical	1787	2144	2207	20	40
West Midlands Sample	2022	2991	3255	48	60

(c) Potential added value

Fig. 1.7 Analysis of potential added values
(Source: Norman and Bahiri (10))

forced to improvise. Over that period, despite the nominal drop of 40 per cent in available capacity, the actual output statistics for manufacturing showed an average drop of only six per cent.

Some well known examples can be cited to support the argument that surplus capacity is plentiful. The famous Fawley productivity agreements for the construction and maintenance of the Fawley refinery (32), showed that much surplus capacity was available to be organised. Overtime earnings amounted to 32 per cent of the wages bill. In exchange for a basic pay rise, overtime was nearly eliminated and many restrictive practices were bargained away. Five and a quarter hours were cut from the working week, but productivity rose by 50 per cent.

Other examples of wasted labour have been recorded. One of the most spectacular was the 1966 shake out in the motor industry when BMC made 10 000 men redundant and maintained the same output levels.

More specific analyses of labour and machinery utilisation have been made by Norman and Bahiri (10). They surveyed companies in the West Midlands and concluded that in terms of potential utilisation levels, taking the best individual firms average as a basis for the industry group, net output or added value per employee could be increased by over 50 per cent. The details are given in Fig. 1.7. On the machine side they found only 41 per cent of the machine capacity was utilised in adding value to materials. A similar study by Swan in

Humberside machine tool companies **(11)** compared a company's highest recorded levels of labour and machine utilisation against its own average and found that the mean potential improvement in labour utilisation was 72 per cent and in machine utilisation was 61 per cent.

Clearly, existing capacity both in terms of labour and equipment right across industry is not being used at anything like its full potential in the UK. It is not only in terms of utilisation levels that there is slack. Stocks, space and the technical limits of equipment could all be used to greater effect. There is in fact a very large potential for improvement in the use of production capacity by British industry.

1.8. RELEASING THE POTENTIAL

It is clear that unexploited physical resources are available. Increases in productivity and output are pointless, however, unless the products can be sold. The measure of success is not so much output per man as sales per man. It is necessary therefore to define what will generate sales in terms of design, price and delivery.

1.8.(a) Design

On the first selling point, design, Britain could well be reasonably competitive. The UK is an acknowledged leader in the fields of inventiveness and design. Figure 1.8 (a) shows that this country supports a high level of research and development (R and D). This should lead to good design capability, but there are doubts if the effort has been directed to the right industrial sectors. In an interesting study of this question, Sir Ieuan Maddock **(12)** correlated the size of each industrial sector with the extent of R and D and compared the UK position with Japan; see Fig. 1.8 (b). He concluded that, 'In the case of Britain, the distribution of R and D does not correlate in any obvious way with the size and significance of the sector involved.'

The eccentric nature of this distribution of research resources is even more striking if that element of R and D funded by the government is highlighted. It can be seen that two relatively small industries, aerospace and electronics, receive most of the government funds, while other very large sectors such as mechanical engineering, food, drink, tobacco and construction receive very little.

	(1)	(2)	(3)	(4)	(5)
	R & D expenditure excluding defence, atomic and space $ million 1963	*Total R & D expenditure as % of GNP in 1964*	*R & D expenditure in col. (1) as % of GNP*	*Growth of real GNP per head % p.a. 1953-65*	*Comparative levels of real consumption per head US=100 1965*
USA	7803	3·4	1·3	1·7	100
UK	1296	2·3	1·4	2·5	69
France	715	1·9	0·9	3·8	69
W. Germany	1173	1·4	1·2	4·9	68
Japan	865	1·4	1·3	8·3	41
Italy	230	n.a.	0·5	4·6	46

(a) *Research and development expenditure*

Net output with amount spent on R and D.

	UK		Japan	
Commodity	*Net output £m*	*R & D £m*	*Net output £m*	*R & D £m*
Aircraft	592	208 (160)	63	7
Chemicals	2070	127	3015	239
Construction	3400	3	5320	34
Electronics	1120	179 (80)	1950	170 (40)
Food, Drink, Tobacco	3058	27	2010	34
Gas, Electricity, Water	2111	6	1230	10
Manufacturing Industries	4226	61 (15)	4060	239 (60)
Mechanical Engineering	2617	54 (25)	3060	94
Metal goods	1380	9	1340	13
Metal Manufacture	1871	21	2420	72
Mining, Quarrying	960	4	505	9
Motor Vehicles	1494	57	1850	116
Other Electrical	930	40	1234	120
Paper & Printing	1837	4	1850	12
Shipbuilding	414	7	708	8
Textiles	1317	13	1590	22

(b) *Comparison of UK (1972) and Japan (1971)*
(*Figures in parentheses show element of R and D funding provided by government*)

Fig. 1.8 Research and development

1.8.(b) Selling price

On the second point, selling price, Britain's prices after a series of devaluations are also competitive on the international market, although this position may soon be destroyed by current inflation levels.

1.8.(c) Delivery

The real British weakness and the point where at present it cannot compete, is in the organisation of the making of products, leading to a disgraceful record for the delivery of products and unnecessarily high costs. Delivery performance is a basic problem which can only be improved by changes in production organisation. In a study in 1966 of the hindrances to exports, 23 per cent of companies identified delivery performance as a major factor preventing increased exports.

Delivery performance is the single most important impression the customer has of the producers internal effectiveness. The pointers are clear. There is spare capacity and there are wasted resources in most companies. Improving production management in order to improve delivery, provides the key to a more competitive company offering better customer service and the prospect of future sales and employment. These in turn will lead naturally to achievement in investment and industrial relations.

1.9. THE PRESSURE OF NEW SOCIAL FORCES

It has been shown that changes in organisation are necessary in order to simplify production and bring back some measure of control to production management. There is also a need for change to meet changes in our social system. Changes in the education system, and several years of affluence have made it difficult for the present generation of workers to accept traditional methods of factory organisation.

1.9.(a) The effect of size

Increases in the size of production organisations have tended to depersonalise workers. The tasks that they do have become more and more fragmented and the imposition of centralised planning and control systems has largely eliminated any contribution that they can make to the planning and control of their own work.

Large organisations are structured hierarchically and tend to be based on the belief that maximum economic return results from designing jobs which minimise training and skill requirements while ensuring full machine loading. Undoubtedly in the past major gains in efficiency came from exploiting the potential of de-skilled and functionalised jobs but now a point of diminishing returns from such an approach has been reached. Such jobs are unattractive to enter and unsatisfying to perform. In consequence planned profits are tending to be offset by high rates of absenteeism and labour turnover, a low concern for quality and increased industrial conflict.

1.9.(b) Changes in the educational system
Changes in the educational system have made jobs in factories less attractive to the younger worker. Stress is now placed in schools on the involvement of the learner, the questioning of given information, and the analysis of the implications of theory and prescribed action. Young people are encouraged to believe that they matter as individuals and that what they have to say is worth listening to. Young workers consequently expect, need, and want, reasonable explanations for what they are asked to do, opportunities for self expression, and a measure of self determination. Taught that they live in a democractic society they cannot blindly accept authority and discipline. Furthermore today's young worker must be seen not as a contemporary phenomenon, but as the vanguard of a changing work force in which the new worker is likely to reject tasks which do not meet personal needs and provide opportunity for individual growth.

Fortunately the problems in the design and organisation of work can be seen in a more positive way. Research indicates that 'work' is still generally considered a worthwhile activity. Whether this fact has direct biological origins or is culturally based is unimportant. The fact is that work gives most people a firm footing in their society and provides essential links with reality. Studies of the retired and of the unemployed demonstrate this fact convincingly.

1.10. THE SOURCE OF INITIATIVE FOR CHANGE
It is clear that change is necessary. It is also clear, as will be shown in this book, that in Group Technology there is a new approach to the

organisation of work which can give some of the types of change which are needed. It is not quite so clear from where the initiative for industrial change should come.

Pressure for change in companies comes from four different sources. Firstly government policies have a direct bearing on industrial behaviour. Secondly, developments within the industry to which a company belongs, such as technological invention and market forces, affect company practices. Thirdly, the company management itself takes initiatives, and finally, individuals either as customers, shareholders, or employees can force companies to adopt different approaches.

1.10.(a) Pressure from individuals
It might be thought that an individual can exert very little pressure on a company. However, the individual is in the end the company's customer. As the purchaser of a company's products or services, the customer should be highly critical of faulty designs, late delivery and high prices, and thus force efficient management back on to the producer. The real competitive position of a company is determined more by pressure from the purchasers of its products than by the competitive image presented by its marketing, sales and public relations specialists.

Shareholders too have more power than they realise. One man gave formal notice of a question to be asked at the annual general meeting of a large corporation, in which he had a few shares. He asked: 'Why in view of our present liquidity problems, are we doing nothing to introduce Group Technology?' He was entertained by the Directors and given a long explanation of why 'it wouldn't work here'. His effort was not immediately rewarded but does illustrate the fact that directors are usually sensitive about suggestions that they are 'leaving stones unturned'. With persistance shareholders can persuade companies to change.

Individuals in their role as employees inside a company have much less scope for initiative in resource management. Unless they are self employed or work in a very small organisation, it is rarely in their interest to make better use of resources. The payment system and the unions can both circumscribe the individuals desire and opportunity to raise standards.

1.10.(b) Initiative from government

Many would say that it is up to the government to provide the lead
in the improved management of resources. This is, however, one
area where the government can do very little. It can create the right
environment and provide information and advice services, but no
government can directly legislate for efficient works management
within companies in a free economy.

The three greatest benefits which the government could offer
industry would be stability of the total economy, consistency of
industrial policy, and provision of efficient advisory and retraining
facilities. There is little evidence to suggest that the government is
succeeding on these fronts. The 'stop-go' government policies which
have persisted since the second world war have prevented management from taking important investment steps and have led to confusion in their employment policies.

The governments failure to provide the right kind of environment
may be because they do not have the right capabilities. Many
alternatives have been tried. Shonfield and Britten relate the fasci-
nating track record of the government's constantly changing policies
on incomes, prices, taxation and growth plans since 1945 (13), (32).
Britten suggests that the pattern of control has probably broken down
because of the sheer complexity of the issues. Professor John Jewkes
has made perhaps the shrewdest suggestion (14). He says that, 'the
moral to be drawn is not highly complicated, nor is it dependent upon
complex and highly sophisticated economic analysis. It is simply that
the cure for bad (government) planning is not better planning, but
no planning'.

Quite apart from questions of competence, governments might well
be questioned on their level of real interest in the efficient running of
factories. Remarkably few civil servants have spent any time in
industry. No government has shown any political awareness of the
works management functions at the level where resources and people
are actually managed.

Certainly the government does not place much obvious store on
meeting its own production targets. In the defence field for example,
there is no record of any government sponsored weapons having been
delivered on time since the war and actual costs have always exceeded
original estimates by a factor of three or more. The resource wastage

in the nationalised industries, over which the government is supposed to exert some direct control, is widely known and publicly ridiculed. In a study of the productivity of the UK government-controlled industries (steel, coal, airlines, transport, etc.) productivity in terms of output per unit of capital employed was found to be universally the lowest in comparison with equivalent European industries.

1.10.(c) The initiative of management
There is little hope that government can or will exercise a detailed, effective and immediate policy for improving resources management. It rests mainly on the managers of companies to make the necessary changes to improve the situation.

The initiative for a new effort in the management of production must come from the top in each factory, because the changes needed affect all management functions. The current failure of our factory management is a consequence of a creeping paralysis which has gradually overtaken our whole system of industrial control. The existing factory managers and production managers certainly intend to do their jobs well but their efforts are frustrated by the forces of the system which has developed over time. Three forces dominate particularly: firstly, the fragmentation of management functions due to increased functional specialisation, secondly, the increased dominance of large companies whose vast organisational problems have caused new management functions to be introduced above the level of the factory manager, and finally the new social forces pressing in on the work force which the production manager is supposed to organise.

An assault must be launched on this whole system for a return to effective resource management and effective customer service through the managers directly in charge of physical operations.

1.11. CONCLUSIONS
This chapter has looked at some of the problems facing the engineering industry in Britain. It has attempted to show that these problems are caused not so much by the environment in which industry works as by the way in which work is organised inside our companies and factories. It has pointed at the extreme complexity induced by excessive process specialisation and by the growth in size of production

units, as the main cause of our present troubles. There does not seem to be any reason why large companies should not be efficient if they practice devolution. What does now appear to be certain is that attempts to control large companies in detail from the centre are generally unsuccessful and that factories and departments which employ very large numbers of workers, tend to be less efficient than those which employ fewer workers.

The research has shown that there is potential for improvement. There are reserves of capacity if the present problems in managing production can be overcome. The research has also indicated that this capacity could be usefully used to increase sales if present problems with deliveries can be solved. The problems of poor delivery arise directly as a result of our failure in production management.

Change is needed in the methods of organising production, not only for economic reasons but also to meet the pressure of new social forces. New entrants to industry have had freedom during their education to question old philosophies and to express their own opinions. They do not take kindly to the extremes of automated centralised control and to the autocratic, bureaucratic methods of supervision which are still used in some factories. Their opposition is manifested by growing rates of absenteeism, and industrial unrest.

The research has indicated that Group Technology can help to solve many of our present problems. The research has also indicated that the source of initiative for change must come from management. The initiative must come first from the boards of directors and senior managers in manufacturing companies. The main aims of this book are first to persuade management to take the initiative and second to give advice on how to do so, which is based on the experience of companies which have already made the necessary changes.

CHAPTER 2

DESIRABLE
CHARACTERISTICS OF GROUPS

2.1. INTRODUCTION

The previous chapter described some of the problems facing the British engineering industry and noted the urgent need for change. It submitted that in Group Technology there is a new method of organisation which can overcome many of our present difficulties.

This chapter now defines Groups and Group Technology. It describes the historical development of the new methods and notes that almost identical methods of organising work in groups have been developed by both production engineers and behavioural scientists. The fact that these new methods of organisation appear to solve both economic and social problems is an important reason for the growing belief in their importance.

This chapter also examines the desirable characteristics of groups and notes that the companies which have been most successful with Group Technology, have been those which have formed groups with these 'desirable characteristics'.

Mention is made of the way in which the change to groups from traditional methods of organisation, affects the working life of both workers on the shop floor and managers. Finally the change to Group Technology is presented as a total company change. Fundamentally it is a change from an organisation of people based mainly on processes, to an organisation based on completed products, components and major completed tasks.

All the research teams involved in the research contributed to the findings reported in this chapter. As was to be expected, they did not agree on all the details of the definitions and desirable characteristics. This chapter attempts to report the ideas expressed and to emphasise the importance of those ideas which had the widest support.

2.2. THE NATURE OF THE CHANGE

The traditional approach to the organisation of production is to use 'line layout' where possible, and 'functional layout' in all other cases.

33

2.2.(a) Line layout

With line layout the machines or other work stations are laid out in a line in their sequence of usage. Usually it can, therefore, only be used efficiently when all components made on the line use the same work stations in the same sequence, and where there is an approximate balance between the work loads at each station.

For these reasons line layout is used mainly in simple process industries, and in assembly, where because most operations are manual, it is fairly simple to divide up the work to achieve an even work load at all stations on the line.

In component processing in the engineering industry, with which this book is mainly concerned, line layout can only be used for parts required in very large quantities, or for families of relatively simple and similar parts, all of which use the same machines in the same sequence.

Line layout is therefore only used in component processing to make a very small proportion of the parts made in industry, and Adam Smith's pin factory can be seen as a rare exception rather than the general rule.

2.2.(b) Functional layout

Where line layout cannot be used, the machines and other work centres are used to make many different parts in batches. The traditional approach to the organisation of batch production is known as functional layout, and is based on process specialisation. The workers in the factory are divided into organisational units each of which specialises in a particular process or part of a process. This book is based on research which has been carried out mainly in engineering work shops. In this case the traditional approach to organisation is to form 'sections' which specialise in such processes as turning, milling, drilling, grinding and gear cutting.

Until recently this traditional approach to factory organisation was universal and unquestioned. Its acceptance has been independent of culture or politics. Factories organised in this way are found in all countries in the world. They are found in capitalist and in communist countries, and they are found in privately owned factories, in state owned enterprises and in industrial co-operatives.

FROM : PROCESS-SPECIALISING SECTIONS

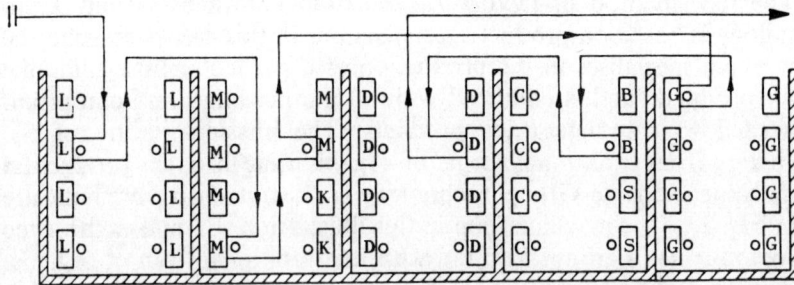

L=lathes; M=millers; D=drills; C=gear cutting; B=broach; S=shaper; G=grinder; K=keyseater.

Each part visits many sections.
Workers specialise in one process only.
Most sections contain only one type of machine.

TO   : GROUPS COMPLETING "FAMILIES" OF COMPONENTS

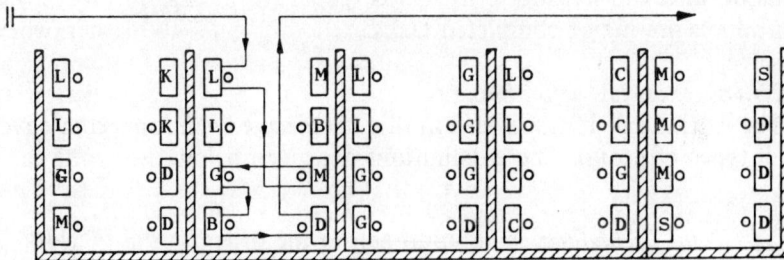

Each part visits only one group.
There is the possibility of choice: some workers can specialise and some can work a variety of machines.
Most groups contain several types of machine.

Fig. 2.1 The change to groups in machine shops

(*This illustration is an extract from* Final Report on a Study into the Effects of Group Production Methods on the Humanisation of Work, *published by the ILO's International Centre for Advanced Technical and Vocational Training, Turin, Italy, and is presented here with their approval as copyright holders*)

2.2.(c) *Group organisation*

The new method of factory organisation known as Group Technology is based on product specialisation. In this case each group of workers specialises in the production of a particular list or 'family' of 'products' and is equipped with all the machines and equipment needed to complete these products. The physical nature of the change from traditional forms of organisation in batch production machine shops, to Group Technology is illustrated diagrammatically in Fig. 2.1. The machine shop in this illustration still makes the same components in groups as it did with the traditional form of organisation. It still uses the same machines and the same methods and tooling. The main change has been in the organisation of the workers and in the layout of the plant.

2.3. DEFINITION

This change to organisation in groups is not found only in component processing work shops. Similar changes have been made in assembly, in service departments, in process industries and in offices. In machine shops the products which form the basis for organisation are components. In assembly departments the products are assemblies, or major stages in assembly. In service departments and in offices the products are major completed tasks.

2.3.(a) *Working definitions*

The working definitions used in this book have been framed to cover all types of group. These definitions are given below.

(1) *Group Technology* is an approach to the organisation of work in which the organisational units are relatively independent groups, each responsibile for the production of a given family of products. The smallest organisational unit is the group, but the same principle of organisation is used when forming larger organisational units such as departments.

(2) *A group* is a combination of a set of workers and a set of machines and/or other facilities laid out in one reserved area, which is designed to complete a specified set of products. The workers in a group share a series of common output targets in terms of lists of products to be

completed by a series of common due-dates. The number of workers in a group is limited by the need to obtain social cohesion.

(3) *A major group* is a larger organisational unit than a group which carries out a large range of compatible processes. It fits the definition of a group with the exception that there is no limitation to the number of workers. Groups can only be formed efficiently inside major groups.

(4) *A family* is the set of products produced by a group. These products take the form, in different types of group, of products as sold, assemblies, major stages in assembly, sets of components, or major tasks to be completed.

The definition given above sees Group Technology as a major innovation in the field of organisation rather than as an innovation in technology or plant layout. This emphasis reflects the idea that the people in a factory are more important than the machines.

2.3.(b) Existing groups

The first reaction of many production engineers to these definitions, will be, 'I thought it was new. We've had some sections organised like that for years.' This is true. Most factories already contain some groups, even though they are not known by this name.

Typical examples are the tool rooms in small and medium sized engineering factories. They contain a team of workers, and are equipped with the range of different machines needed to complete their family of products. They normally contain only a small number of workers.

Many assembly sections have the same characteristics and are in effect groups, even if they are laid out as assembly lines. A mass production car assembly line employing say, 450 workers, is not a group, because it has too many workers. A small assembly line with say, ten workers can be 'a group, laid out for line flow'.

Finally most component processing lines employ few workers, have all the other characteristics of groups, and are in fact groups in all but name.

2.3.(c) Synonyms

The concept of Group Technology has evolved rapidly in the last thirty years and has changed radically in the process. The name is not a good description of the concept as it exists today. For this reason several attempts have been made to change the name. It isn't only the French who confuse names with descriptions.

One synonym supported in some academic circles is the term 'cellular manufacture'. This is based on the biological analogy with natural cells. It has not been used in this book for three reasons. Firstly, Group Technology is much the most widely used term and is the term which has been used in most of the literature on the subject. Secondly, the word cell has unfortunate connotations. It tends to remind people of prisons, mental asylums, the communist party, terrorists and espionage. Finally, previous attempts to rejuvenate important concepts by renaming them, have generally failed to have the desired effect.

A second synonym is 'group production methods'. This is used by some international organisations as a general term to cover all organisation in groups in all types of technological system. They use the term Group Technology to cover only component processing groups.

In a few companies where the main aim has been to achieve social benefits, the total concept is known as Job Enrichment or Job Structuring, although both these terms have wider connotations. Finally there are some companies which have invented their own names for groups. They have called them islands, mini factories and work teams amongst other things.

2.4. HISTORICAL DEVELOPMENT OF GROUPS

The use of groups in production organisation has two main origins. Firstly, these methods were developed by engineers who were mainly interested in finding methods which would reduce stocks and work in progress, reduce throughput times and reduce setting time, thereby increasing capacity. Secondly, organisation in groups was developed by behavioural scientists who were looking for ways to increase workers motivation and job satisfaction.

2.4.(a) The engineers development of groups

The development of Group Technology by engineers sprang initially from the work of Professor Mitrofanov of Leningrad University **(15)**.

He found that considerable reductions in setting time and therefore increases in capacity, could be achieved with lathes if similar parts were loaded on the machines one after the other. He also demonstrated that major savings in tooling costs could be effected by these means.

The early work on Group Technology was devoted to exploiting these findings. Without any change in organisation, attempts were made to plan the sequencing of work on machines in order to reduce setting time and achieve an increase in capacity. These early examples of Group Technology are usually referred to as the Single Machine Approach. The single machine approach gave no savings in stocks or throughput time and was found difficult to apply efficiently in any sections other than those which did first operations.

The next development was to place supporting machines such as milling machines and drilling machines next to the lathes to form groups composed of a mixture of different types of machine, which completed components. These early groups were usually formed by selecting parts by eye from the floor of the shop. A few families of components were found which were similar in shape and could be machined on the same groups of machines. It was believed at this stage that groups would only be suitable for a limited range of parts. This stage in the development of Group Technology is sometimes called the Pilot Group stage. These pilot groups demonstrated that major savings were possible but their method of selection tended to complicate attempts to form further groups.

The next stage was the development of methods for planning a total division of work shops into groups. Most of the early methods were based on classification and coding methods, but these have now been largely replaced by production flow analysis. There are now a considerable number of companies with work shops totally divided into groups.

This experience with total division soon demonstrated that the division into groups at work shop level and the change to group layout were not enough on their own. To achieve major successes it was necessary to change the departmental organisation in many companies and also to change many of the supporting systems. In particular changes in production control, in payment methods, in production planning and in methods of supervision were found to be desirable in different cases. This has led to the 'Total System Approach' to Group Technology, which not only forms groups at work shop level but changes the rest of the organisation and support systems as necessary to support them.

2.4.(b) The behavioural scientist's approach to groups
The behavioural scientists were looking for new methods of work organisation in order to increase workers motivation and job satisfaction. Unlike the engineers who were mainly interested in machine shops, most of the work by behavioural scientists has been done in assembly, in process industries, and in offices.

Early pioneers in this work were the Tavistock Institute of Human Relations who formed highly successful groups in a coal mine, in a textile factory in India **(18)**, and in many different industries in Norway. Since then the main initiative in industries other than component processing, has been outside Britain. It is interesting to note that Britain has at present a lead in the number of applications in component processing, which it is rapidly loosing to the USA, but that the rest of Europe – Sweden, Norway, Holland, France and Italy in particular – has more applications in assembly, process industries, and offices, than we do.

The groups created by engineers and behavioural scientists have very similar characteristics. Both fit the definition of a group given earlier in this chapter. Both the engineers and the behavioural scientists have succeeded in most of their applications of groups in achieving the benefits for which they were looking. This fact, that organisation in groups can lead to both economic and social advantages, is responsible for the strong belief held by most of those who have worked with these methods, that Group Technology provides a solution for many of the main problems now faced by industry.

2.5. DESIRABLE CHARACTERISTICS OF GROUPS
The six main characteristics of groups given in the definition were that there should be a set of workers special to the group; that there should be a set of machines and equipment which is special to the group; that these machines and equipment should be laid out in one special area reserved for the group; that the group should complete its own special set of products; that tasks and output targets should

be given to the group as a whole and not separately to the different individuals in the groups; and finally that the number of workers in the group should be small enough to obtain social cohesion.

These six characteristics are those which give Group Technology its special benefits. In effect they divide a machine shop into a number of smaller independent machine shops.

2.5.(a) *Advantages of groups*

It is difficult to describe the desirable characteristics of groups without relating them to the benefits which they induce. These benefits are discussed in detail in Chapter 4. Here only the major advantages are listed as an aid to the present discussion.

By centralising the authority and responsibility for making complete parts in the groups, the following major advantages may be obtained:

(1) because all operations on each part are done inside one group, throughput times can be greatly reduced;

(2) the reduction in throughput times reduce the stock of work-in-progress in the group;

(3) because all operations to make each part are carried out inside the same group, the responsibility for quality can be assigned efficiently to the group with a consequent improvement in quality levels;

(4) for the same reason the responsibility for completion of work by due-date can also be assigned to the group, with a consequent reduction in overdue orders;

(5) provided that parts are ordered together on the groups in sets at regular intervals, it is possible to plan the sequence of loading operations on some types of machine, in order to obtain a significant reduction in setting time and an increase in capacity;

(6) because the needs for component production co-ordination are peculiar to each group, centralized co-ordination is unnecessary and much of the day-to-day planning and co-ordination of the work in the groups can be delegated to the groups; this can give both economic and social advantages;

(7) Group Technology provides a 'climate' where job satisfaction can grow.

To obtain these advantages of groups, it is necessary that they should have the characteristics given in the definition. These are therefore the 'desirable characteristics' of groups, because without them the full advantages cannot be obtained. Unfortunately there are situations were it is difficult to achieve all these characteristics in their pure form. These cases will be discussed for each characteristic in turn in the following paragraphs. In addition to the six main characteristics from the definition, there are other characteristics which have been found from experience to be desirable. These are also discussed below.

2.5.(b) The workers in the group
The provision in the definition that each group should have its own special team of workers, infers that they should remain together in the group and should not be transferred from one group to another. This has the advantages that they learn to work together as a team, and that they learn to understand the special problems associated with their machines and equipment and with the family of components which they make. It also has the advantage that it provides close social contact between workers who are contributing to the achievement of common objectives.

Any change in this characteristic is undesirable. There are however, certain situations where some movement of workers between groups, or 'labour mobility' cannot be avoided. In the case of jobbing production for example, it may be difficult to maintain an even load of work for all the groups. Some workers may have to be moved occasionally from lightly loaded groups to heavily loaded groups. Again in the case of an influenza epidemic, if one group loses most of its workers it may be necessary to transfer some workers from other groups to keep production going until they return.

To summarize, it is highly desirable to maintain this characteristic of workers association with 'their' group. Workers should not be moved from one group to another for trivial reasons, but there may be emergencies when it cannot be avoided.

2.5.(c) *The machines in the group*

The provision in the definition that the groups should be equipped with all the machines and equipment they need to complete the products in their families, is again highly desirable. This characteristic infers that parts should not move out of a group for operations on machines in some other group and that parts from other groups should not come into the group for special operations.

This characteristic is desirable because it is necessary in order to achieve minimum throughput time, work in progress, and handling costs, and because it is unlikely that groups will accept the responsibility for quality and for completion of work by due-date, if they do not control all the processes involved.

Unfortunately, this characteristic cannot always be achieved in all groups. A typical example in machine shops arises when there are heat treatment operations in the middle of series of machining operations. For safety reasons it is undesirable to install such heat treatment equipment as cyanide baths and batch type furnaces in the same groups with machine tools. This means that some parts will have to go to a heat treatment group for the intermediate operations. One company has overcome this problem by treating part of its heat treatment department as a service department for the two groups which require intermediate heat treatment operations. They have sufficient capacity to give a guaranteed throughput time of 24 hr for any intermediate heat treatment operation. The two groups can therefore still schedule their work efficiently.

Once again then, although this characteristic is highly desirable, there are occasions when some variations cannot be avoided. What should be avoided at all costs is any attempt to move work operations from one group to another in an attempt to keep certain types of machine fully loaded. If the groups have been well designed and the output target has been correctly calculated to use the group capacity efficiently, attempts to transfer operations from one group to another, will only cause confusion and can lead to labour difficulties because the men in the groups can no longer have any certainty about the load of work which is being assigned to them.

2.5.(d) *One reserved area for the group*

This characteristic is an almost unavoidable characteristic, if a group is to be called a group. The research has shown that on the very few occasions where companies have attempted to form groups without changing the plant layout, they have achieved very few of the potential advantages of Group Technology. They have failed, for example, to obtain significant reductions in throughput time, to reduce stocks, to engender any team spirit among the widely dispersed members of the group and have been unable to introduce any effective participation in decision making.

Not only should the machines of a group be laid out in one reserved area, they should also be laid out as far as possible to minimize handling and maximize social contacts. The technique of 'line analysis' can be used to analyse the flow between the different machines in a group so that the best layout can be achieved **(16)**.

2.5.(e) *The set of products produced by the group*
The provision in the definition that each group should have its own set of products which it always makes, is the key characteristic of groups. It has the advantages that throughput times and work in progress are reduced; quality is improved; more reliable delivery by due-date can be achieved; all the tools and methods data can be stored in the groups, and that the workers in the groups become accustomed to the special problems of the production of their family of parts.

This characteristic infers that parts should not be moved freely from one group to another. If the groups have been correctly designed, in fact, it will be difficult to move complete parts from one group to another. Generally only the simpler parts with very few operations on common types of machine, can be made in several different groups. The majority of parts will be tied to the groups where any special machines required to make them are installed.

The number of parts in each group family is generally dictated by the nature of the product. In companies making only a small variety of simple products in large quantities, the number of different parts in each family will be very small. With complex products made in great variety, each group may have a large number of parts in its family and they will be of many different types.

2.5.(f) *A common output target*
This characteristic is essential if full advantage is to be taken of the

use of a planned sequence of loading to reduce setting times and throughput times and to increase machine capacity. These savings depend on planning the sequence of loading operations on machines, so that parts with similar set-ups are loaded one after the other. This is only possible if the similar parts are all ordered together at the same time intervals.

The same types of saving can also be obtained with functional layout, providing again that similar parts are ordered together. With Group Technology, however, the reduction in throughput times makes it possible to work with high batch frequencies. This increases the number of set-ups, and makes it even more important to achieve a reduction in the time per set-up.

Ideally output targets should be given to the groups at regular period intervals. For example, every two weeks each group might receive its output target for the next two weeks and this would be accompanied by a summary of the load imposed by the required output, giving the machine hours load on each machine in the group. The materials for each period would ideally be delivered to the group before the start of every period.

This characteristic has the advantage that it simplifies production control, that it makes it possible to plan the sequence of loading on the machines to minimise set up time and increase capacity, and that because the calculation of period load figures is very simple, the workers in the group can assure themselves that they are not being given an excessive load. The fact that the load is assigned to the group as a whole and not separately to each individual in the group, makes it possible to delegate the details of work scheduling to the groups.

In effect this desirable characteristic infers the need for single cycle ordering systems such as Period Batch Control. In the few cases where this type of system cannot be used, it may be difficult to realise fully the advantages of this characteristic.

2.5.(g) The number of workers in the group should be small

This characteristic of groups must of necessity be somewhat imprecise. It is a subject on which there is still very little consensus of opinion.

Behavioural scientists have in the past preferred small numbers of workers per group. Examples are the six to ten range preferred by the behavioural scientists who planned the groups at Philips Television factory in Holland **(17)** and A. K. Rice's recommendation **(18)** that groups should contain either two or six to twelve workers. He excluded three, four and five workers on the grounds that these sizes of group tend to throw up outcasts. The problem that arises if one tries to fix specific limits for group size inside these ranges, is that they would exclude some of the most efficient Group Technology applications now working.

Some production engineers on the other hand have tended to favour larger sizes of group. Large groups can be formed in existing machine shops without the need to purchase any additional machines. Very small groups tend to require an additional investment in machine tools. Again machine utilisation is generally better with larger groups. The balance of load on the groups, imposed by different product mix requirements, tends to be more even from period to period with larger groups. Finally supervision costs are lower with larger groups when, as is usually desirable, supervisors are assigned each to a particular group.

In some cases technological and design features may make it essential to form large groups. For example a company making some very complicated parts which require say 40 operations on 30 different types of machine may require, if it is to form an economical group which completes the components, say, forty different machine tools and thirty or more workers in the group.

One machine tool company, Alfred Herberts Ltd, is running a highly successful and cohesive group with 35 workers in it. It is probably desirable that groups should be smaller than this for most purposes, but experience with Group Technology, is still too limited to fix definite limits.

A postal survey by the London Business School which obtained 30 replies, found that most companies reported the average number of workers per group in the two to ten range. However physical checks in four of these companies found that they had all underestimated the average.

The detailed studies made by Bradford and Salford Universities, in the four companies described in Chapter 4, had average workers per group of 9·5, 15, 15 and 13, and groups varying in size from two to 52.

2.5.(h) Supervision of groups

Characteristics of groups must now be considered which are not covered by the definition, but have been found by the research to be desirable. One of these is that the supervisors responsible for groups should be assigned to particular groups and should each have one group only.

The research has found companies which use foremen as functional specialists, one looking after all the lathes in all groups, another looking after all the milling machines in all groups and so on. This arrangement has tended to eliminate many of the possible advantages of groups. The most successful solution is one foreman, chargehand or group leader per group.

2.5.(i) Participation

One of the advantages of Group Technology is that it makes possible the efficient delegation of responsibility for the operation of the group to the group itself. In most companies this delegation will be to the foreman of the group. There are advantages to be obtained, however, if he can be persuaded to delegate to the workers in his group some of the decisions concerning the operation of the group.

The types of task which have been successfully delegated to groups, include planning the sequencing of jobs to reduce setting time, scheduling the flow of work between machines, deciding which workers in the group should work which machines, inspection and quality assurance, setting up of the machines, and the planning of production methods.

In addition to delegating decision making concerning the way in which the work should be done, some companies have also delegated a measure of social responsibility to their groups. Groups can be found which have a say in the setting of manning levels, in the taking on of new employees and in the manning of shifts. If the load of work allocated to the group in one period is higher than usual, one solution is to allow a certain number of overtime hours to the group. The decision who should do overtime and when, is again one which has been left, in some companies, to the groups to decide for themselves.

Some group participation is always desirable for both economic and social reasons, but not all these types of delegation will be possible or desirable in all companies. In some cases special training may be necessary. In most cases the best method of introducing 'shop floor participation' is to let it grow gradually rather than to introduce it by edict.

2.5.(j) Organisation and control characteristics

Even with groups which have all of these desirable characteristics, the results achieved will depend to a great extent on the way in which the groups are integrated into a total company structure. The planning and control methods used with traditional forms of organisation are often unsuitable for use with groups. Many will have to be changed.

Production planning methods will need changing, for example, to ensure that new parts are planned for completion inside groups. The introduction of new machines and the development of new methods will involve either the modification of existing groups, or in some cases, possibly the design of new groups.

The production control methods in common use in traditionally organised factories will again need changing. Stock control methods of ordering based on re-order levels, in particular, are ineffective when used with groups. The tendency to delegate responsibility to the groups for scheduling the work on their machines will greatly simplify production control and will involve some simplification of existing procedures. This problem is considered in greater detail in chapter 8.

Payment methods may also require some modification. In general, individual financial incentive schemes are not ideal for use with Group Technology, because they involve the use of additional, otherwise unnecessary, paperwork, and because they make it difficult to achieve efficient flexibility of labour. Systems which are more appropriate and have been used successfully with groups, include group bonus, straight day work and monthly salary. Of the four companies investigated in detail in Chapter 4, two use day work payment, but the other two still use individual incentive payments.

Other systems which may need changing include product design, costing, marketing, particularly in relation to sales forecasting, and personnel management.

2.5.(k) Social characteristics of groups

One of the advantages claimed for Group Technology was that it

Desirable characteristics		Social factor
1. Team	Set of workers special to group.	Feeling of belonging.
2. Facilities	Set of machines and equipment special to group.	Association with the means of production.
3. Group Layout	One area special to the group.	Association with a territory. (Territorial imperative).
4. Products	Set of products made only in the group.	Association with the completion of a significant product.
5. Target	Output targets set for group as a whole.	Common objective and simple Feedback necessary for achievement motivation.
6. Size	Small number of workers.	Group stability and cohesion.

Fig. 2.2 Desirable characteristics of groups

provides a 'climate' where job satisfaction can grow. The desirable characteristics of groups, which have been jusified above mainly in terms of their operational and economic benefits, also have important social implications.

Figure 2.2 now shows the desirable characteristics of groups, together with their related social factors (19). Sociological and psychological research has shown that each of these factors has an important effect on job satisfaction. Group Technology with these desirable characteristics can therefore provide conditions conducive to an increase in job satisfaction. The desirable characteristic of groups are desirable for both economic and social reasons.

2.6. EFFECT OF THE CHANGE TO GROUPS

The change to Group Technology will inevitably affect working conditions in the factory. It is important that management should understand the nature of these changes because the most difficult job in introducing Group Technology is to persuade people to accept new ways of working which are different to those to which they have been accustomed throughout their working lives.

2.6.(a) Effect on the worker

The worker in the traditionally organised machine shop will have been used to working in a 'section' containing machines all of the same type. Each worker has his own individual task assigned to him and as far as the work is concerned, there is little need for co-operation between the different workers in the section, because they don't generally work on the same components. Normally the worker will spend most of his working life operating one machine or one or two similar machines.

With Group Technology he now co-operates with the others in his group in order to complete components. Perfect balance between machine capacity and load has never been possible even with traditional methods of organisation. There are in most factories therefore, more machines than men. This ratio may be quite small in mass production factories making a small range of components, but may be two, or more, to one in jobbing machine shops. This fact of life is repeated in most groups, but in this case the worker instead of transferring when necessary from one machine to another machine of the same type, may on occasions be asked to change between machines of completely different types. The term generally used to describe this type of change is 'worker flexibility'.

In practice it has been found that the fact that some machines are heavily loaded while others are lightly loaded, allows an element of choice to the workers in the groups. Those who prefer to stay mainly on one machine can do so if they operate one of the heavily loaded machines. Those who prefer a variety of work can move between the machines which are lightly loaded.

For most of the workers there will be little change in the number of different machining operations which they do. Many of the different machine types in the factory will exist in one group only. It is probable that the men who used to operate these machines in the traditional sections will move with them into the groups which now make the types of components which are processed on them. It is only the workers who work common types of machines such as lathes and drilling machines, for example, which are found in more than one group, who may suffer some reduction in the variety of different operations which they do.

The worker in a group will have the satisfaction of working in a team which completes components. There is considerable sociological and psychological evidence that this association with the completion of products is an important factor in job satisfaction.

In traditional forms of organisation where materials have to travel between many different sections before they are completed, the main task of management is co-ordination of processing needs and of material flow. It is difficult with these traditional systems to see how one could hope to achieve good quality and delivery by due-date without strong central control. With groups all the co-ordination needs for each family of parts are concerned with only one group. In this case it is not only possible but economically desirable to delegate the work of co-ordination to the groups themselves. Once again there is considerable evidence that this closer association of the foremen and the workers with the management of their own jobs, can have a significant effect on job satisfaction.

2.6.(b) Effect on the foreman

In many traditionally organised work shops the foremen have little real authority. In the organisation of work they are expected to follow procedures, plans and methods which are imposed upon them from above. They carry the final responsibility in a situation where they have little possibility of doing anything to improve matters.

Most companies which have changed to groups have appointed foremen or other supervisors to run them. These men find themselves running what is in effect an independent machine shop. Some have found the new responsibilities too much for them, but for the majority the change to groups has been very welcome.

The problems for the foremen only come later if management try to persuade them to delegate some of the decision making to the workers in their groups, or the workers in the groups themselves, having acquired a good knowledge of the problems in the group, start to question the orders of the foreman and to try to influence decisions. For foremen of the old school who have been used to obedience without question, this change can be difficult.

It seems probable that in the future the nature of the foreman's job will change. He will have to become less of an autocratic supervisor and more a councillor-motivator and technical expert in the group. This points to a need for changes in the job descriptions, selection methods and training for supervisors.

2.6.(c) Effect on the line manager

The line managers who run the different production departments in the factory will also find that their work has changed. Where today much of their work deals with co-ordinating the flow of materials between the different sections they control, with routine data processing tasks, and with the settlement of incentive payment disputes, with groups there will be very little of these types of work. They will be more concerned with setting objectives for their groups, controlling to see that they are achieved, with training and with the development of new methods.

Once again it will be difficult for some line managers used to autocratic control to change their approach, but for most of them the change will be welcome as it makes their work more professionally interesting and easier to control.

It will be seen that with groups there is a tendency for decision making to be delegated to lower levels in the organisation. The line manager has to delegate some of his old responsibilities to the foremen in charge of the groups and they in turn are expected to delegate some of their decisions to the workers under them. Some authorities have suggested that the best way of introducing Group Technology is not to start with groups on the floor of the shop, but to work from the top down. The change of organisation in a big company would start with a division into main product divisions with much wider decision making responsibilities delegated to them. They in turn would be expected to divide their divisions into independent produce-based major groups or departments such as machine shops, forges, and assembly departments and to delegate a large measure of decision making to these departments. Finally the departments would be divided into groups and the line managers of the departments would be expected and encouraged to delegate some decision making to their groups.

2.6.(d) Effect on the staff manager

The managers and clerical workers engaged in staff functions, are likely to be the most affected by the change to groups. One cannot delegate a measure of decision making to the line management and supervisors without changing in some measure the work done at present by the staff specialists.

Such jobs as ordering, dispatching and progressing in production control, for example, are likely to be simpler and to require fewer people in the future than they have with traditional systems. Payment systems and costing methods again can become simpler and may require fewer staff. It is important that management should understand this long term trend, so that they can reduce new employment in these fields and allow natural wastage and retraining to effect the necessary reduction in manning levels.

2.7. CONCLUSIONS

This chapter has defined groups and Group Technology and has given a brief account of the historical development of the new approach.

It has shown that groups have two origins. Very similar groups have been developed by production engineers, with mainly operational and economic objectives, and by behavioural scientists with mainly social objectives. The fact that groups appear to help solve both economic and social problems has been an important factor in their development.

The chapter has examined the desirable characteristics of groups, and has shown that they are desirable for both economic and social reasons. The research on which this book is based has indicated very clearly that the companies which have been most successful with Group Technology have been those which have formed groups with these desirable characteristics.

The chapter has also studied the way in which the change from traditional methods of operation to Group Technology is likely to affect the worker, the foreman, the line manager and the staff manager and clerical staff.

CHAPTER 3

GROUPS IN THE CONTEXT
OF USER FIRMS

3.1. INTRODUCTION

This chapter examines what has already been done by companies in the engineering industry in Britain, to introduce Group Technology. It is based on studies made by the London Graduate School of Business, and by Bradford and Salford Universities. The research shows that Group Technology has been applied in a wide range of companies making many different products and using many different production systems. It is difficult today to find any type of engineering product which has not already been made in groups somewhere in the world.

A comparison between the development of Group Technology in Britain and in other countries shows that Britain in 1975 had more applications than any other country and that, in particular, in the application of groups to component processing in the engineering industry, there were more examples of groups in Britain than in any other country. This lead is, however, being eroded rapidly, particularly by the USA. Group Technology is still in the early stages of development; less than four in every 1000 of the companies in Britain have attempted to use it.

An examination is made of the motives which have induced companies to introduce Group Technology and also of the reasons why a number of companies, which started to plan for Group Technology, decided to abandon the project.

A list of companies which have introduced at least one group, is included at Appendix C, and a list is also given in Fig. 3.5 of completed applications of Group Technology. An analysis is made of the problems with Group Technology which are being experienced by some of these companies. Detailed studies of Group Technology in four of these companies are included in the following chapter.

3.2. LIST OF APPLICATIONS

A list of sixty seven applications of Group Technology in the engineering industry in Britain is given at Appendix C. These companies make

	No. of companies	Out of replies	%	PI	IC
Process.					
Assembly	9	67	14	4	6
Machining	63		97	43	20
Sheet Metal	3		3	1	0
Office	1		1·5	0	1
Progress.					
Partly installed (PI)	41	67	66		
Instal: complete (IC)	25		34		
% in Groups					
100%	10	26	38		
51— 99%	0				
10— 50%	10		38		
1— 9%	6		23		
Used pilot groups?					
Yes	16	55	29		
No	39		71		

(a) *Details of groups (Total of 67 companies)*
   (*Note: some companies have used Group Technology for more than one process*)

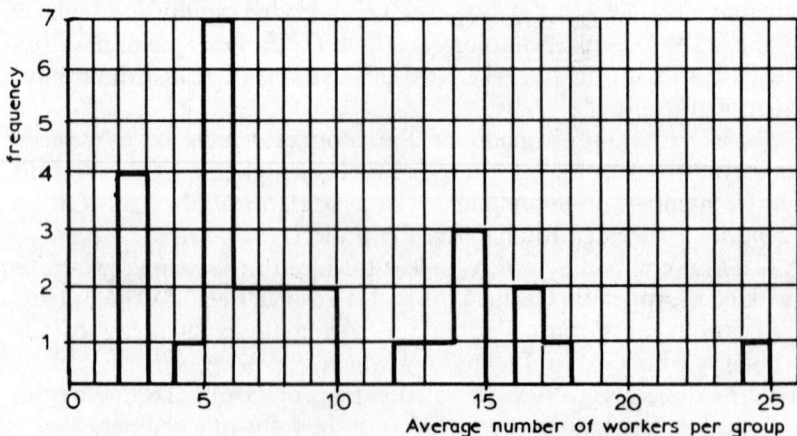

(b) *Workers per group (30 replies)*

Fig. 3.1 UK engineering applications of Group Technology

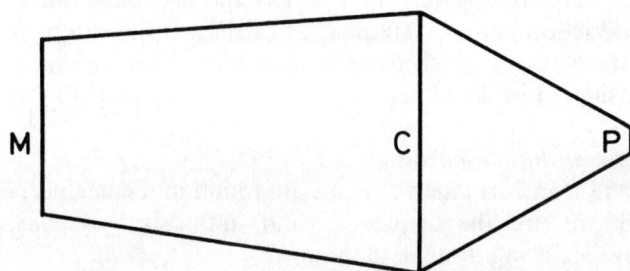

(a) *Explosive systems. Large number of material items and bought parts. Trans-formed into a larger number of components, used to make a small number of assembled products.*
Examples: Ferranti, Alfred Herbert, Serck Audco

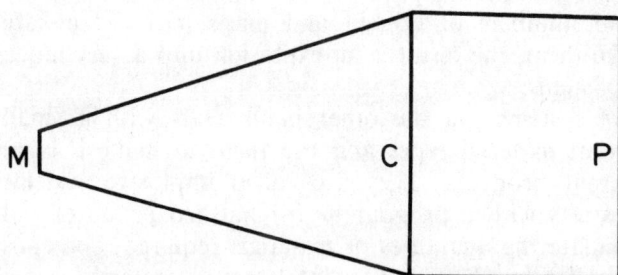

(b) *Implosive systems. Small number of material atems used to make a large number of components. Mainly sold as components, but a few sold in sets or as simple assemblies.*
Examples: Dormer, Ferrodo, BOC

> Key: *M=number of different bought material items and parts*
> *C =number of different components*
> *P =number of different products*

Fig. 3.2 Explosive and implosive systems

a wide variety of different engineering products and use many different types of production system. An analysis of these applications is given in Fig. 3.1. Ways in which these applications differ one from another are considered in detail below.

3.2.(a) Types of transformation system

Two main types of transformation systems are found in the engineering industry. There are the 'explosive' and 'implosive' systems, illustrated in Fig. 3.2 using 'funnel' diagrams.

The explosive systems are those which start with a wide variety of materials and bought parts, convert the materials into a larger number of finished parts and then combine these parts together into assemblies for sale. The majority of the companies listed at Appendix C are of this type. They are called explosive systems because if one starts with the production programme for products and calculates the amount of materials and numbers of bought and made parts which are required to make them, the result is an explosion into a very much larger number of items.

The implosive systems, on the other hand, start with a small variety of different material types and use them to make a large variety of different products. They are called implosive systems because if one starts with a programme for finished products and attempts to calculate the quantities of materials required, one ends with a much smaller number of different types of material. Most jobbing foundries and many jobbing machine shops are of this type. Some of the best applications of Group Technology in Britain are in implosive industries. The research indicates that the type of transformation system does not effect the feasibility of using Group Technology.

3.2.(b) The commercial system

There are three main types of commercial system found in the engineering industry. Different companies either make to order, make standard products for stock, or make standard products against a forecast of future sales. Examples of all three of these types are found among the companies listed in Appendix C, together with different combinations of all three systems. The research indicates that the type of commercial system does not affect the feasibility of using Group Technology. There are successful applications using all three of these commercial types.

3.2.(c) Product differences
There is a wide variety of different types of engineering product included in Appendix C. These products vary in the precision to which they have to be made, the quantities of each type which are made per year, in the varieties of each product type which are made and in the types of industry which use them. Coupling British experience with experience in other countries, it is found that there is scarcely any type of engineering product which is not now being made at least partly in groups.

3.2.(d) Plant type
The types of plant found in engineering groups is also highly varied. It ranges from the very small machines used in instrument manufacture, to very heavy machine tools used in the manufacture of large engineering products.

The plant also varies greatly in sophistication. Groups can be found using old semi-manual types of machine and at the other extreme there are groups with highly sophisticated numerically controlled machines and groups which are fully automated in processing, material transfer and inspection.

It appears that Group Technology can be applied successfully with any type of plant or equipment.

3.2.(e) Type of layout
Groups can be found in the list of applications given in Appendix C, which are laid out both as lines, and as small general engineering groups in which different components use different combinations of machines in different sequences. Usually it is only the groups which make simple types of product which have a consistent processing sequence, that can be laid out as lines.

The research indicates that Group Technology can be used efficiently for both simple products where line flow is possible inside the groups and for very complex products where this is impossible.

3.2.(f) Size of production unit
Finally the production units which have introduced Group Technology have varied from a small jobbing shop with 43 direct operatives, to a large company with over 5000 workers on the site. These
production units have varied from single privately owned enterprises,
to single factories in large corporations employing 20 000 or more
people on different sites in many different geographical locations. It
will be seen that the size of the company does not appear to restrict
the feasibility of using Group Technology.

In general, experience with Group Technology to date indicates
that there are no limitations to its use in the engineering industry. It
can be, and has been, used to make nearly every conceivable type of
engineering product in almost every conceivable type of production
system.

3.3. AN INTERNATIONAL COMPARISON
It is interesting to compare the British achievement in introducing
Group Technology, with that in other countries in the world.
Burbidge carried out a study in 1965 into the effects of group production methods on the humanisation of work **(19)**. In the course of
this study he found 477 different applications of group production
methods in 32 different countries. This list covered all types of
industry and included 25 companies which were in the planning
stages of introduction and did not yet have groups in production.

3.3.(a) Component processing
As far as component processing is concerned, Britain in 1975 had
more applications of Group Technology than any other country and
accounted for nearly one third of the total in this field. Most of these
groups are in machine shops but there are examples in forges, sheet
metal shops, and foundries.

Four applications of component processing groups from outside
Britain complement British experience and are worth special mention.

The first of these is the application of Group Technology in the
factory of Soc. Guilliet at Auxerre in France. This company makes a
range of wood working machine tools and other products. It started
by divisionalisation, and then proceeded to divide its production
divisions into groups. Group Technology is now company wide. This

application is of particular interest due to the high participation by the work force in planning the change. Mr Mary, the Chairman and Managing Director, has developed in this company the idea of Project Planning Groups. Temporary multi-disciplinary groups, with members from different levels in the organisation and different functions, have been used to plan many of the changes made.

A second application is the use of groups at the factory of RIV SKF at Airasca in Italy. This company makes ball bearings and is very highly automated. Fully automated machining lines and a connected assembly line produce and pack ball bearings into cartons, almost without human intervention. The groups of people who manage each line are in this case mainly concerned with the maintenance and servicing of the line. They carry out such duties as keeping the feed hoppers filled with balls, seals, and cages, checking the accuracy of the fully automated inspection equipment, changing the setting of the grinding machines on the machining line in order to maintain the right balance of race sizes for selective assembly, and minor maintenance on the conveyors and other equipment on the line. This application is important because it demonstrates that automation changes the nature of the work done by the workers in a factory, and tends to encourage the formation of semi-autonomous groups to manage the automated lines.

Another application of Group Technology can be found in the Stork Werkspoor factory in Holland which makes very heavy diesel engines for marine purposes. This application is of particular importance because it contains groups with perhaps the heaviest machine tools which have yet been moved for group layout. The cost of moving such large machines is often considered prohibitive. After careful study this company believed that the change in layout was worthwhile.

3.3.(b) Groups in assembly
In the case of assembly, Britain has a few groups in production, but nothing to compare with the groups which can now be found on the mainland of Europe. Well known examples of the use of groups in assembly can be found in the car assembly plant of Volvo at Kalmar in Sweden, in the Olivetti factory at Scarmagno near Ivrea in Italy, where they are used to assemble electronic calculators and tele-

printers, and in the Philips factory at Eindhoven in Holland, where groups have been used for a number of years for the production of television sets.

3.3.(c) Other types of groups

Britain is also behind in the application of other types of group. Groups are now being used in the organisation of process industries, offices and services. Among well known examples in these types of occupation, the Bell Telephone Company in USA has changed to group organisation for the production of telephone directories. There is a famous application of group organisation in a company making pet foods in Topeka, Kansas, USA. The Norsk-Hydro chemical factory in Norway is also mainly organised in groups. Finally there are a number of applications of group organisation in insurance offices and in banking in several different countries.

3.4. REASONS FOR INTRODUCING GROUP TECHNOLOGY

One part of the study undertaken by the London Business School, attempted to discover the reasons why companies either introduced Group Technology or decided not to carry on with it. Fifty-four different companies were contacted, of which 38 had implemented Group Technology, and 16 had started but then abandoned the project.

3.4.(a) Reasons for implementation

The reasons for introducing Group Technology given by the 38 companies which had succeeded in introducing at least one or two groups, are analysed in Fig. 3.3. These responses were obtained partly during visits, partly by post, and partly on the telephone. All the companies gave lists of different reasons for introducing groups. The first eight of the reasons given in Fig. 3.4 are particularly relevant since all were expressed by three or more companies.

The single most stated reason for group implementation was the desire to speed throughput time and the second most important reason was to reduce work in progress. It is evident that the latter can follow from the former. The first eight most frequently stated reasons are all concerned with a desire to increase production performance and this has been the main motive for implementing Group Technology in Britain.

Reason	Number of times stated
1. To speed throughput	22
2. To reduce work in progress levels	11
3. To reduce stocks for cost and space saving	7
4. To improve deliveries	7
5. To improve control lost through increased size	6
6. To improve productivity	6
7. To reduce set-up times on automatic machinery	4
8. To increase capacity	3
9. To enrich jobs	2
10. To improve quality	2
11. To maintain profitability in each market sector	2
12. To improve standardisation of design	2
13. To eliminate operations where possible	1
14. To cope with low output as result of small batches of many components	1
15. To overcome problems of shortages and rushing through small batches of urgent parts	1
16. To make each marketing production unit separate	1
17. To make sensible cost centres	1
18. To rationalise production	1
19. To improve labour relations	1
20. To reduce labour turnover	1
21. To improve departmental performance	1
22. To check on work in progress	1
23. To locate parts needed urgently	1
24. To attribute responsibility	1
25. To provide new planning systems in area of low volume components	1
26. To give young managers more responsibility	1
27. To rationalise, simplify and clarify management	1
28. To speed assembly	1

Fig. 3.3 Reasons for introducing Group Technology

None of the stated reasons refer directly to customer relations. It is surprising in view of the poor delivery record of the manufacturing industry, that only 20 per cent of the companies gave improved

Reason	Number of companies stating it
1. Difficulty with coding. Got as far as component classification systems	4
2. Changes in company structure or management—company divided into several subsidiaries, change of director, change of personnel at production management level	3
3. Impossibility of reaching agreement with the labour force—Trade Union antagonism, disagreement on payment systems	2
4. Wrong type of industry, e.g. continuous process industry in which group bonus scheme or job enrichment schemes applicable but not cells	2
5. Bought large machine to do everything and scrapped cells	1
6. Economic recession	1
7. Informally practising the use of cellular manufacture but no decisive action, i.e. 'rough form' practised and discussed but no implementation with the driving force of the whole company involved	1
8. Had unsuccessful trial in one area only—difficulty in isolating cell and in changing work of some personnel 'overnight'	1
9. A design for cellular manufacture in the company was the subject of an unfinished Ph.D thesis. As the thesis was never concluded, the design was not implemented	1

Fig. 3.4 Reasons for failure to implement Group Technology

delivery performance as a reason for adopting Group Technology. As many of the companies, however, make products only on receipt of customers orders, some of those who were anxious to improve throughput times, may have been concerned with the underlying effect of long throughput times on delivery performance.

Only four companies gave reasons for introducing Group Technology which were related to labour relations. These were expressed as: 'to aid labour relations'; 'to enrich jobs'; and 'to reduce labour turnover'. None of these reasons was expressed in isolation but were given in lists of reasons, indicating an awareness of the multi-faceted benefits obtainable from the use of Group Technology and a desire to obtain all these benefits.

The companies which gave reasons for adopting Group Technology, were asked whether these reasons had been justified and if any additional benefits had been noted. It appeared from their replies that whatever the expectations of a company in relation to Group Technology, these were generally realised. In other words, if a company introduces groups to solve a delivery problem, the groups will improve deliveries although this one benefit may actually be part of a wider complex of changes, sometimes unmeasured, which are brought about by Group Technology.

Only one company described the benefits it had achieved as 'adequate' in comparison with all the other companies who referred to them as 'good' or 'very good'. All the companies reported a substantial reduction in throughput time and increased productivity. To summarise, there are many stated reasons for the adoption of Group Technology, most aimed at the improvement of productivity, control and management. These more tangible improvements carry more weight in decision making than the more qualitative social changes, although both aspects are generally considered.

3.4.(b) Reasons for not implementing Group Technology

The sixteen companies which failed to implement Group Technology, gave nine reasons for not continuing with the change. These reasons are listed in Fig. 3.4. These answers were given by companies which had considered, discussed and even in some cases designed the groups for their companies, but had failed to implement Group Technology for various reasons.

It appears that there are two common features in all these failures:

(1) the lack of a determined individual within the company pushing the implementation of groups;
(2) the lack of a company wide understanding of the nature of

Group Technology and its possible benefits, leading to a consequent lack of confidence.

With certain reasons, such as where increased mechanisation had been introduced, the failure to implement Group Technology may have been reasonable. Several of the reasons stated, however, represent no more than a failure to present the concept of Group Technology in an acceptable way to all concerned, a lack of long term planning, or an inability to cope with the changing industrial situation. It is difficult for example, to understand how 'difficulties with coding' can be such an insurmountable problem and doubtful if it is even relevant.

3.5. COMPLETED APPLICATIONS

Turning again to Appendix C and the list of engineering companies in Britain which have introduced Group Technology, it will be noted that 26 of these companies have described their implementation as complete (IC). These companies are listed in Fig. 3.5. In most of these cases the company had limited objectives, but in all of them they have succeeded in completely changing the layout and organisation in some major department into the forms associated with Group Technology. In most of these companies there were other departments which had not yet changed to groups.

3.5.(a) The best applications

About ten of these applications have achieved major economic successes by the change. These benefits will be considered in detail in Chapter 5. It is evident that the companies which have had the best results from the use of Group Technology are those which have most nearly followed the desirable characteristics of groups described in the previous chapter. It is probably true to say, however, that even among the first ten best applications there are few which exhibit in their best form, all these desirable characteristics. In other words examples can be found of companies which have formed groups with many of these desirable characteristics, but few have succeeded in achieving all of them.

All the 26 companies in Fig. 3.5 claim some benefits but some of them have failed to achieve the major gains which are possible. This

1. *Baxendale Ltd*, Bamber Bridge. Baxi Gas Fires.
2. *Bentley Engineering Ltd*, Knitting machines. Leicester.
3. *BOC Ltd*, Bletchley. Welding nozzles.
4. *BOC (Welding) Ltd*, Edmonton.
5. *BOC Ltd*, Skelmersdale. Welding equipment.
6. *Chubb Lock & Safe*, London. Safes.
7. *Chubb Lock & Safe*, Wolverhampton. Locks.
8. *Ferodo Ltd*, Chapel-en-le-Frith. Brake Linings.
9. *Ferranti Ltd*, Edinburgh. Air borne radar.
10. *V & E Friedland Ltd*, Stockport. Door chimes.
11. *Garrick Engineering Ltd*. Jobbing machining.
12. *H. J. Godwins Ltd*, Pumps. Cirencester.
13. *Herbert m/c Toold Ltd*. Coventry.
14. *International Engineering Ltd* (ICL), Belfast, N.I.
15. *Jones & Shipman*, Small tools. Leicester.
16. *George Kent Ltd*, Luton, Beds.
17. *Mather & Platt Ltd*, Manchester. Electric motors.
18. *Thomas Mercer Ltd*, St Albans. Measuring instruments.
19. *Newage Engineering Co*, Coventry. Gearboxes.
20. *Platt Saco Ltd*, Bolton. Gears.
21. *Rank Xerox*, Mitcheldene, Glos. Copying machines.
22. *Serck Audco*, Newport, Shropshire. Valves.
23. *Simon Container Machinery Ltd*, Stockport.
24. *Simon VK Ltd*, Wirral. Printing machines.
25. *Westinghouse Brake & Signal Co. Ltd*, Chippenham.
26. *Whitaker Hall Ltd*, Manchester. Fluidair Compressors.

Fig. 3.5 Engineering applications of Group Technology in UK reported 'completed'

is not surprising because these companies are the pioneers of a new innovation, and in many cases they have been obliged to develop the new methods, with little knowledge of the methods used, and problems encountered, in other companies. Most of these companies are now working to develop their groups in order to obtain the full advantages of Group Technology.

3.5.(b) Main deficiencies in existing applications
It will help at this stage to consider what were the principal problems experienced by these early applications. The detailed studies of applications were limited to only seven companies. The problems

discovered in these companies have, however, since been found to be typical of those in other applications.

First of all, several companies are having problems due to the design of their groups. For efficient group production it is essential that they should exhibit all or most of the desirable characteristics of groups. The capacities of the groups should also be approximately in balance. In other words there should be sufficient capacity in every group to produce the components needed to meet the programmed requirement of finished products. Some companies with groups which were not well designed, are having difficulties in obtaining an even load of work on the groups and in completing the components ordered on them, by due-date. They have found temporary solutions such as transferring certain operations to other groups, or moving workers from one group to another, but these solutions tend to eliminate important potential advantages of Group Technology.

A second main problem has been concerned with information. It is essential for efficient Group Technology that the information data used for planning and control should be changed to recognise the existance of groups and should also be maintained at a high level of accuracy.

A third common problem found in several existing applications of Group Technology is with the production control system. Companies which have attempted to use existing multi-cycle stock control systems to control production in groups have found such problems as a highly variable load of work on the groups from period to period, and difficulties in obtaining the reduction in setting time which is possible with a planned sequence of loading.

Purchasing also has been a problem in some companies. Even with Group Technology, one cannot make products without materials. It appears to be desirable when introducing Group Technology, to improve the material supply system first, before continuing with the change to groups.

Another common problem is concerned with the lack of understanding among the personnel in companies, of the nature of Group Technology. This indicates a lack in training of the personnel. Without this understanding some companies have introduced very efficient Group Technology layouts but have failed to get the best from these installations because they have made changes during operation, which eliminated many of the desirable characteristics of groups.

3.6. CONCLUSIONS

The research has indicated that groups are being used to make a wide variety of engineering products by a wide variety of different methods.

Compared with other nations, Britain leads in the number of applications in engineering machine shops, but does not have as many examples of the application of group production methods in assembly, in offices and in process industries. Group Technology is, however, still at a very early stage of development. Less than 0·35 per cent of British engineering companies have attempted the innovation.

The main motive for introducing Group Technology in British factories has been a desire to simplify production and improve its performance. This preference supports the thesis of this report, which sees deficiencies in production management as the main cause of our present industrial problems.

There are about 26 applications of Group Technology in Britain which might be said, in a limited sense, to be complete. About ten of them have achieved major economic advantages. There is still a great deal to be done, however, before most of these 'complete' applications can achieve the full advantages of Group Technology. There has been a tendency in the past to believe that once groups were installed on the floor of the shop, the maximum benefits would come automatically. Now that more is known about Group Technology ways can be seen in which most of the applications could be improved.

It might be said that there are still two main jobs to be done in Group Technology. The first is to persuade more companies to make the change and the second is to develop the existing applications to obtain their full possible advantages.

CHAPTER 4

FOUR EXAMPLES OF GROUP TECHNOLOGY

4.1. INTRODUCTION

Detailed studies of Group Technology applications in a number of companies were made initially by Manchester University, UMIST, and later by Bradford and Salford Universities. Their study reports are here condensed into four case studies.

To make these cases comparable, each in divided into sections under the same headings.

The factory.
The products.
The production processes.
Where Group Technology is used.
Motives.
Constitution of groups.
Supervision.
Payment method.
Production control.
Parts mobility.

Labour mobility and flexibility.
Delegation to groups.
Costing.
Design.
Production planning.
Benefits achieved.
Method of introduction.
Date of introduction.
Changes in layout.
Conclusions.

The four companies which co-operated in the preparation of these condensed reports were Ferranti Ltd, Edinburgh, BOC Ltd, Bletchley, Platt Saco Lowell Ltd, Bolton, and Whitaker Hall Ltd, Radcliffe.

4.2. FERRANTI LTD, CREWE TOLL, EDINBURGH

4.2.(a) The factory
The Scottish Group of the Ferranti company comprises five main centres employing 5800 people. The Crewe Toll factory in Edinburgh is the largest of these centres, employing 5000 people.

A central machine shop situated at Crewe Toll serves all the other factories in the group. About 350 people are employed in the machine shop.

Group	Type of component	No. men	No. m/cs	No. parts	Batch quantity			Ops. per part			Total No. of Ops
					Min.	Max.	Av.	Min.	Max.	Av.	
A	Hubs and pivots	2	9	7	200	3000		6	9		
B	Simple turned parts over 2in dia.	15	13	1200	1	300		2	10	7	8000
C	Complex castings	14	17	515	1	281		4	20	17	8886
D	Spigot gears	10	15	1198	5	900		2	8	8	9740
E	Flat gears	8	5	393	1	200		6	12	9	3453
F	Non-round milled and drilled parts	8	10	1242	1	200		2	6	9	11786
G	Auto turned parts	12	14	1419	100	20000		2	6	5	8296
H	Non-metalic flat plates	3	3	1000	1	300		3	5	4	4000
J	NC milled parts above 35mm	8	8	229	1	270		7	12		3004
K	M/C shop dalkeith										
L	Complex turned and ground parts	15	17	546	7	300		4	20	9	5196
M	NC milled parts under 35mm	8	15	1282	1	270		7	12		16433
N	Lab workshop										
P	Sheet metal pressed parts	20	17	2352	1	200		3	20	6	14206
Q	Complex sheet metal components	8	4	2356	1	40		3	8	5	11587
T	Simple turned parts below 1in dia.	7	12	2323	5	2000		2	7	5	11290
X	Items not falling into other groups	23	52	2961	1	300		2	40	6	18826

Average 15

Fig. 4.1 Composition of Ferranti groups

4.2.(b) The products

The factory makes avionic equipment including airborne radar and inertial navigation sets. These are high precision products with much of the work requiring very close tolerances.

The main customer is the British Government. Contracts are normally placed for 200 to 300 sets at a time, with deliveries spread over three years.

4.2.(c) The production processes

Assembly and test are the main production processes. The Crewe Toll factory buys most of the components used in its products, but has its own machine shop and sheet metal workshop.

The level of technology in production is high. In the machine shop, the company pioneered the introduction of tool pre-setting, digital read-out and numerical control (NC). It now has a high proportion of NC machines and is a pioneer of computer aided design and manufacture.

4.2.(d) Where group technology is used
Fourteen groups are installed in the central machine shop. These cover all the direct labour and all the machine tools in the department. There are three groups in the sheet metal department, one of which covers experimental work; again all the workers and machines in this department are in groups.

4.2.(e) Motives for introducing groups
The machined parts represent only 7 per cent of the works cost of the products, on average. Reduction in the cost and stock investment for machined parts was not, therefore, a primary motive for introducing Group Technology. If, however, any of the machined parts are late, a substantial investment in stocks of bought out items can be induced.

Prior to the introduction of Group Technology, delays in the completion of machined parts were a major cause of delays in assembly. The primary aim was therefore to reduce the throughput times for machined components and so make them available at the right times to permit efficient assembly of the whole product and thus minimise the total investment in stocks and work in progress.

4.2.(f) Constitution of groups
The numbers of men, machines and components, in each group and the types of component in each family, are shown in Fig. 4.1. Drawings of typical components made in the main production groups are shown in Fig. 4.2.

Excluding Miscellaneous, Fitting and Assembly, Sheet Metal Working, and Services plant, and considering only the production machine shop plant at Crewe Toll, there are 83 different machine type code numbers in the plant list, covering a total of 263 machines. Fifty nine of the code numbers, or 71 per cent, each exist in one group only. The remainder exist in more than one group.

These figures do not include groups doing sheet metal work, Group K, which is at Dalkeith, or Group N (L/W) which makes experimental parts.

4.2.(g) Supervision
Each group has its own cell leader appointed by the management.

CAM CONTROL
Drg. No. A/95/13703

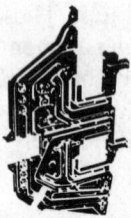

R. F. BLOCK (BOTTOM HALF)
Drg. No. E/95/14242

HOUSING
Drg. No. B/3954/58329

A   Miscellaneous Milled Parts with NC Machining Content. i.e. Continuous Path Milling and Point to Point Milling and Boring.

CASTING
Drg. No.
D/3992/70626

END PLATE
Drg. No. C/3954/702...

FRONT PANEL
Drg. No. D/3954/55315

C   Complex Shapes mainly Castings requiring combinations of Milling, Drilling, Jig-Boring, and Tapping.

SCREW
Drg. No.
A/3854/02161

PIN
Drg. No. A/93/23081

LINER
Drg. No. A/3954/58274

SHAFT
Drg. No. A/3992/70337

G   Turned Parts up to 1" Dia. including High Quantity Very Small Diameter Parts. Centerless Grinding Facility included.

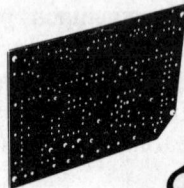

BOARD (BOTTOM)
Drg. No. C/95/14344

TAGBOARD
Drg. No. B/95/14271

H   Flat Plates and Fibre Glass boards High Speed Production including N.C. Drilling and Punching.

RETAINER, PACKING
Drg. No. A/95/13726

COVER
Drg. No D/3992/71062

Q   Sheet Metal Parts requiring Machining mainly Milling.

COVER
Drg. No. B/3992/70167

INNER SKIN ASSY.
Drg. No E/3854/03253

R   Sheet Metal Fabrications and Assemblies requiring Hand Forming and Benchw...

SPUR GEAR
Drg. No.
B/3854/00596

WASHER, CONVEX
Drg. No.
A/95/14668

EAR
g. No.
A/95/12070

GEAR
Drg. No.
B/3992/71268

WASHER
Drg No.
A/3854/02986

SPUR GEAR
Drg. No.
B/3392/70319

SPUR GEAR
Drg. No.
B/3854/00563

SPUR GEAR
Drg. No.
B/3954/57544

CLAMP, RIM CLENCHING
Drg. No A/95/14116

BLOCK
Drg. No. A/3954/71369

PLATE
Drg. No. A/3854/02288

CRANK ROD
Drg No. B/3954/56012

E Disc type Turned Parts mainly comprising Spur Gears produced from Sheet Material.

ned Parts up to 2˝ Dia. with Complex
lling and Gear Cutting facilities.

F Non – Round parts requiring mainly conventional Milling & Drilling.

SWITCH COVER
Drg. No. B/94/12028

S/METAL SEAL PLATE
o. C/3854/02536

SINK
No. A/3992/71514

CAP
Drg. No. B/3954/56064

PLATE
Drg. No. D/3992/71310

CLAMP
Drg. No.
A/95/25303

PLATE
Drg. No.
A/95/04003

OUTER SKIN FRONT
Drg No. C/3854/03217

AIR FILTER 5335-99-113-2865
Drg. No. A/3954/58318

Complex Turned and Ground Parts
including Clutch Parts.

M Aluminium Alloy Milled components suitable for N.C. Milling, Drilling, and Jig – Boring.

P Sheet Metal Parts requiring Guillotine and Power Press Tools.

DOOR SUB-ASSY.
Drg. No. B/3954/71184

TYPICAL COMPONENT RANGE

T Simple Turned Parts 1˝ Dia. and less, requiring simple Milling, Drilling and Tapping. Sequence Technology applied at the first operation.

CYLINDER
Drg. No. C/3854/03298

AZIMUTH GIMBAL
Drg.No.E/3854/00367

X Parts not displaying similarities suitable for the other Machining Groups

Fig. 4.2 Typical components, Ferranti

4.2.(h) Payment method

An individual incentive payment method is used, based on standard times fixed by work study.

4.2.(i) Production control

A production programme is prepared showing product completions necessary to meet promised deliveries. It is based on two week periods. It is revised if delivery requirements change.

Ordering on the groups and on suppliers is based on explosion from the production programme. The ordering method is multi-cycle period batch control, based on two week periods. Each group has a standard batch throughput time and the batch quantities ordered are based on this standard time.

The details of load scheduling are left to the group leaders. To assist them in this work they are provided with computer print-outs at regular intervals showing:

(1) the parts ready for machining on each machine in the group, in priority sequence;
(2) the calculated earliest and latest starting dates for all operations, for the parts on order.

The balance of load in man-hours and machine-hours, between the groups has changed in the long term with changes in design and with the acquisition of new plant. This has necessitated some changes in group manning levels. In the short term the production control system provides an even balance of load between the groups.

4.2.(j) Parts mobility

All parts are allocated to their own particular family and group. It is very rare for a part to be transferred from one group to another.

4.2.(k) Labour mobility and flexibility

Labour mobility is low. Workers are allocated to particular groups. They are only transferred to other groups in an emergency.

Labour flexibility is high. Most of the workers can operate two or more machines in their group, but in practice the more highly skilled operators are reluctant to carry out work at a lower level of skill than their highest level.

4.2.(i) Delegation to groups

The following activities are delegated to the groups, nominally to the group leader:

(1) Setting-up.
(2) First-off and some final inspections.
(3) Detailed load scheduling, including changes in the computer loading priority sequence, when necessary due to machine breakdown, tool breakdown, or absenteeism.
(4) Distribution of tasks to workers in the group.
(5) Some say in new recruitment to the group.

4.2.(m) Costing

A conventional total costing system is used, based on operation costs, and overhead allocation. There has been little change in the costing system since the introduction of Group Technology.

4.2.(n) Design

In the case of parts made on conventional machines, there has been no attempt to alter the design of the parts to suit Group Technology. The materials and tolerances were optimised prior to the introduction of Group Technology. Such optimisation is a continuing practice which is not affected by the organisation in groups.

The picture is different with NC machines. The design has been altered radically in this case to suit manufacture by NC. However, the same change in design philosphy would have given benefits with either functional layout or Group Technology. In other words, machine grouping has not affected the design function. The design has been modified to suit the machining process and not the organisation.

4.2.(o) Production planning

The first task of the production planner now, when introducing a new part, is to allocate it to the appropriate group. Decision flow charts have been introduced to assist in this task.

In some of the groups using conventional machines, a planner is allocated to the group. This means he is a specialist and becomes more effective in his specialisation.

In the case of groups with NC machines, efforts are being made to combine production planning and NC programming with the drawing office tasks, to assist in the introduction of computer aided design and manufacture.

4.2.(p) Benefits achieved
The company has achieved its main aim of a more reliable delivery of machined parts to assembly.

As the company was involved in a major investment programme in NC machines and was making significant changes in design philosophy at the same time as the introduction of Group Technology, it is difficult to apportion the benefits to the various changes which were made. However, some of the benefits obtained were mainly due to the introduction of Group Technology. In particular:

(1) hold ups of partially assembled units were reduced;
(2) customer deliveries were improved;
(3) setting times were reduced in some groups.

Increased job satisfaction and motivation were not the aims of any of the changes. The main change in this respect is that the group leader now has a much clearer job than previously.

4.2.(q) Method of introduction
The introduction of Group Technology was directed by a production engineer reporting to a production manager. Outside assistance was provided by a team from Manchester Univeristy (UMIST).

The first group was found by simply looking at the parts made on a large number of turning machines, using bars of 1 in diameter and under. Some of these parts had simple secondary operations such as milling and drilling. The group was set up empirically.

The second group was for machined castings. It was set up in the same way. Similar groups for the manufacture of parts made principally on NC machines were also formed at the same time. The remaining groups were established by means of Production Flow Analysis (PFA).

The change was preceded by a considerable effort in method and tooling development. For example, digital read-out, pre-set tooling and sequencing were developed.

There was a large investment in NC machines, which were included in groups. This investment reduced the number of machines in the machine shop due to the high productivity of the NC machines.

4.2.(r) Date of introduction
The first group was installed in February 1967. The last group started work in April 1973.

4.2.(s) Changes in layout
The division into groups has proved to be stable. As new machines have become available the make up of the groups has been revised. Design changes have reduced the requirements for gears and complex castings and have reduced the loads on these groups. However, no new groups have been required.

4.2.(t) Conclusions
This is a successful application of Group Technology. It is important because it shows that Group Technology can be used in companies with very advanced technology, a high rate of technological change and with relatively small batch quantities.

4.3. BOC LTD, BLETCHLEY

4.3.(a) The factory
This is one factory belonging to the engineering division of the large British Oxygen Company Limited. The factory has 150 employees. It was built on a new site in 1971.

4.3.(b) The products
The factory makes a range of nozzles for gas welding, flame cutting and heating purposes. The current range of nozzles consists of 134 types. Including specials and other types of nozzle which may occasionally be ordered on the factory, the total range is approximately 500 different types of nozzle.

4.3.(c) The production processes
The transformation process is implosive. Ten diameters of copper bar plus 10 diameters of brass bar are transformed into many products.

The main production process is machining. This includes turning on automatics, swaging, deep hole drilling, milling, diamond turning and flame testing. Unusually for an engineering company, drilling is the process which requires the highest precision.

4.3.(d) Where Group Technology is used

There are eight machining groups in the machine shop. These are supplemented by a small three man group doing flame testing of the finished products and a metal treatment installation operated by one man.

These groups cover all the direct labour and all the machine tools in the department.

4.3.(e) Motives for introducing groups

The company give as their reason for introducing Group Technology a wish 'to produce the highest quality product at the lowest possible cost'. They see this as the essential basis for other desirable economic and social benefits.

More than most companies in Britain this company sees a need for improved management–worker relations and is active in promoting better information services and a high level of workers participation. It sees Group Technology as a desirable form of organisation to help promote this type of change.

4.3.(f) Constitution of groups

The number of men, machines and components in each group and the types of component in each family are given in Fig. 4.3. Illustrations of the types of product made in each group are given in Fig. 4.4.

It will be noted that each group specialises in a limited range of nozzles. In general each of these ranges requires a different type of plant so that this type of division was inevitable.

There are 100 different types of machine installed in the machine shop. Ninety per cent of these types exist in one group only, the remainder exist in more than one group.

4.3.(g) Supervision

There are four 'group leaders' appointed by management and reporting to the machine shop superintendent. The group leaders are all

Group No.	Type of component	No. men	No. m/cs	No. parts	Batch quantity			Ops/part		
					Min.	Max.	Av.	Min.	Max.	Av.
170	ANM nozzle	4+13	24	6	1000	5000	3300	1	8	7
171	PNMR nozzle	2+10	15	5	1000	1000	1000	2	5	4
172	Swaged tip	1+8	13	13	1000	1000	1000	1	4	3
173	Contact tip	1+1	6	22	100	5000	1000	1	1	1
174										
175	PF type nozzle	3+9	32	35	150	500	200	1	4	3
176	AFN nozzle +									
	6290 and 2890	1+4	18	15	500	1000	750	1	5	3
177	AGNM nozzle +									
	some AFN and									
	ASNM	2+7	17	10	500	1600	1000	6	6	6
178	SCH nozzle +									
	process nozzle	3+7	20	27	150	1000	600	5	6	6
	Average:	9·5								

Fig. 4.3 Composition of BOC groups
Note: In the third column, 4+13 (for example) indicates 4 setters and 13 semi-skilled

technical specialists. All of them can set up and operate any machine in their group. Under the agreement with the union they are allowed to set up and operate any machine when they wish.

4.3.(h) Payment method
All workers in the factory are paid day work. There are two rates of pay, one for skilled workers and another for semi-skilled workers. There are no time clocks in the factory but all employees, including the general manager, sign on when they arrive at work and sign off when they leave. The method of payment is by credit transfer to the employees bank account and is payable monthly in arrears.

4.3.(i) Production control
The factory supplies nozzles for five BOC trading units and numerous overseas outlets. Some despatches are made direct to a few of the overseas outlets but most of the output of the factory goes to a central warehouse which attempts to maintain six weeks stock of all lines.

Production control starts with market forecasts from the five BOC trading units. These are received once a quarter or every thirteen weeks. These forecasts are consolidated onto a thirteen week programme which shows the output required of each type of nozzle from each group.

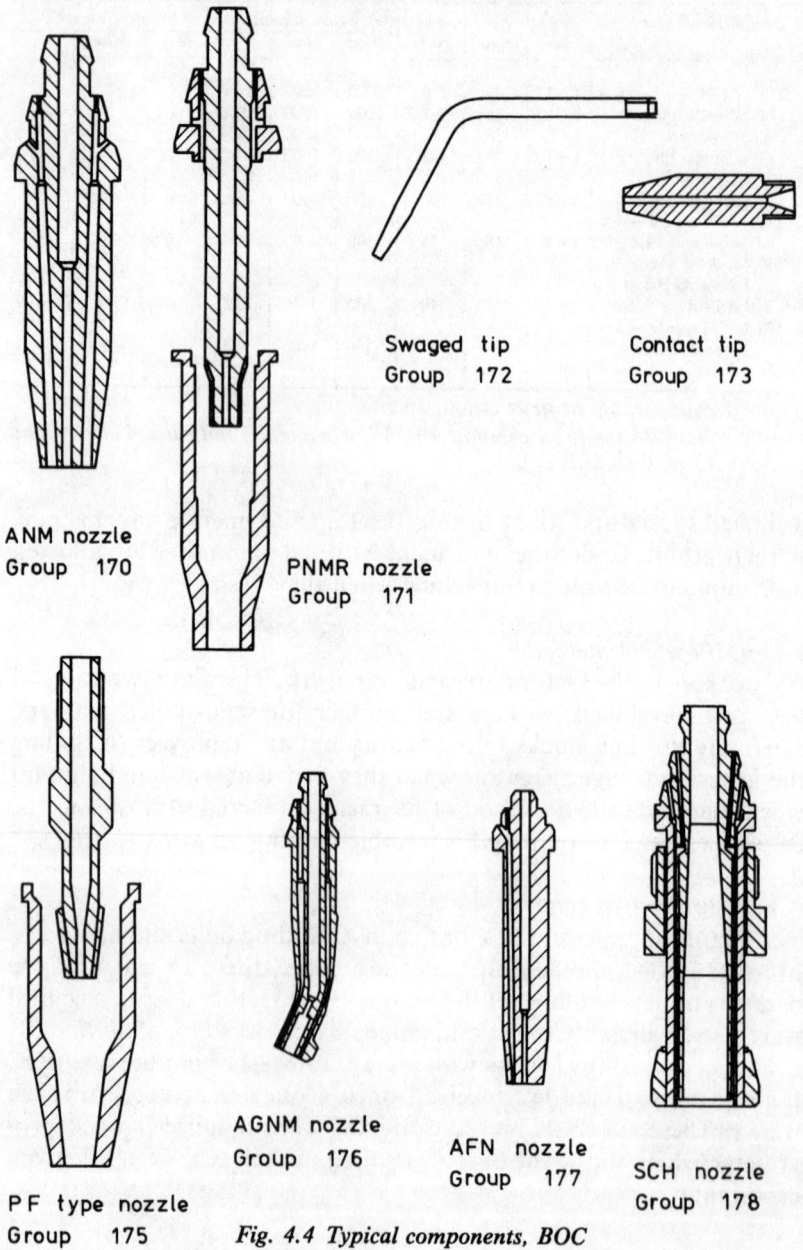

ANM nozzle
Group   170

PNMR nozzle
Group   171

Swaged tip
Group 172

Contact tip
Group 173

PF type nozzle
Group   175

AGNM nozzle
Group   176

AFN nozzle
Group   177

SCH nozzle
Group 178

Fig. 4.4 Typical components, BOC

Once every six weeks a firm programme, or order, is issued to each group showing the weekly output of each type of nozzle required from the group. These six week programmes or orders are prepared at meetings usually chaired by the commercial manager and attended by the group leaders and the production controller.

Apart from the six week programmes the only other paper work consists of a stores requisition and a job card for each batch.

The production control staff consists of one man and one girl.

Pareto analysis was used based on annual output value of nozzles to determine the cycles for the manufacture of different types of nozzle. The Class A high value nozzles representing about 70 per cent of the output are manufactured every three weeks. The Class B or intermediate value nozzles are manufactured every thirteen weeks and the Class C low value items are manufactured to a six monthly cycle.

4.3.(j) Parts mobility
All parts are allocated to their own particular group. There is however some parts mobility between groups particularly between groups 177 and 178. This arises due to irregularity in the demand for the types of nozzle made in these two groups.

4.3.(k) Labour mobility and flexibility
Labour mobility between groups is fairly high, particularly during the holiday period of July, August and September when most of the workers choose to take their holidays. This mobility is recognised as necessary but is not always popular with the workers who have developed a strong sense of group identity.

Labour flexibility on the other hand, or the ability of workers to move from one machine to another inside their groups, is very high. Most of the skilled workers can set up or operate any of the machines in their group which require a skilled operator. And most of the semi-skilled workers can operate any machine which requires semi-skilled workers.

The company has only one union, the AEWU, and their agreement with this union allows complete mobility and flexibility inside the skill ranges. It also allows the group leader to set up or operate any machine in the group.

4.3.(i) Delegation to groups
There is a high level of delegation and participation in the groups. The workers in the group do their own setting and store many of their own tools. They are also responsible for their own inspection. Four inspectors responsible to the quality control engineer also do patrol inspection.

The group leader participates in the framing of his six weekly production programme. He is then completely responsible during the six week period for completing that programme. He allocates the work to the machines and workers in the group, or in other words does his own operation scheduling.

4.3.(m) Costing
The company operate a standard costing system. Each group has its own allocation rate.

In addition to the stores requisition and job card already mentioned, the workers also maintain a time sheet showing their times on different operations. This time sheet is used only by the cost office. The company are at present considering a change in this system towards either group costing or process costing.

The costing system is operated by one man and one woman.

4.3.(n) Design
Most of the nozzles are designed by a development engineer. The change to Group Technology had no effect on the design function.

4.3.(o) Production planning (production engineering)
The planning of production methods is centralised. Only five people work in this section which is, in effect, already a small group.

4.3.(p) Benefits achieved
As this installation of Group Technology was introduced on a 'green site', it is difficult to quote benefits achieved. The same nozzles were however made in a different factory before they were moved to Bletchley and the following benefits are claimed by the company.

(1) A major reduction in work in progress. This used to be 15 to 18 weeks in the old factory and is now only three weeks.

(2) A greatly improved ability to respond to market changes. It is difficult to quantify this benefit.
(3) The management believe that groups have given them greater management–worker harmony than was ever possible with traditional methods in the old factory.
(4) The simpler material flow system has greatly simplified production control. This is now run by one man and one girl and there are no shop clerks and very little paper work in the factory.
(5) The change to groups has also simplified the costing system which again is now operated by one man and one girl.

4.3.(q) Method of introduction
The change to groups was planned by the Engineering Manager with the assistance of an engineer who was later transferred to another factory in the company. No outside assistance was employed. The groups were found by a study of the production processes used to make the different designs of nozzle. The factory started with the machines laid out in groups in September 1971 and has never used any other method.

4.3.(r) Date of introduction
The factory started operation in September 1971 with group layout already installed.

4.3.(s) Changes in layout
The basic division into groups has required very little change. There has however been some combination of small groups and some redistribution of blank turning on automatic lathes.

Group 174 was eliminated by combination into group 175. It is now intended to add group 173 to group 175 as well. This was done mainly due to the difficulty of maintaining an even load of work on very small groups such as 173 and 174.

Initially all the automatic lathes were installed in group 175 which turned the blanks for all the other groups as well as itself. With the acquisition of more modern bar automatics in group 175 some of the old automatic lathes have now been transferred to other groups to make them self contained. In some cases these lathes are left permanently set for one type of blank. In the very near future group 175 will only make its own blanks plus some blanks for groups 173 and 171.

Apart from these changes the only other changes made in layout have been those necessary when new items of plant were purchased and installed.

4.3.(t) Conclusions

This is a well established and efficient application of Group Technology. It is particularly interesting because it demonstrates that Group Technology can be used efficiently in implosive industries most of whose products are single components.

This application is also of interest because it is one of several in BOC. Group Technology has been adopted in this company, particularly in the engineering division, as a policy, which includes establishment of small production units and strong devolution of responsibility. It demonstrates the fact that Group Technology is most likely to succeed when everyone in an organisation, from the Board of Directors down, is determined that it shall succeed.

This is also an interesting application because it is one of the few applications in Britain where the company have strongly supported worker participation at shop floor level.

4.4. PLATT SACO LOWELL LTD, BOLTON

4.4.(a) The plant

The plant in Bolton is one in a large company with five factories. It employs 1000 people. Among other departments it contains the central foundry, gear machining centre, and sheet metal departments for the group.

4.4.(b) The products

The plant makes bottle cleaning and pasteurising equipment and textile machinery. It also makes grey iron castings, gears and sheet metal components for all the factories in the group.

The products made in the gear department include pulleys, ratchets, sprockets and integrally geared and splined shafts as well as a variety of gears.

The products made in the sheet metal department range from large cabinets and covers to small pressings.

4.4.(c) The production processes

The main production processes are metal forming, fabrication, machining, sheet metal working and assembly.

The machining process in the gear centre generally consists of two turning operations; one gear cutting operation and one or two finishing operations.

The processes in the sheet metal department include work on guillotines, press brakes, power presses, and welding.

4.4.(d) Where Group Technology is used

Eight groups are installed in the gear machining department and there are sixteen groups in the sheet metal department. These cover all the direct labour and machines in both workshops.

4.4.(e) Motives for introducing groups

The main motive was, 'to improve the control of component processing'.

4.4.(f) Constitution of groups

The numbers of men, machines and components in each group and the types of component in each family are shown in Fig. 4.5. Typical components are illustrated in Fig. 4.6.

(1) There are 104 machines of 80 different machine types in the gear centre plant list of which 78 per cent each exist in only one group, the remainder being found in more than one group.
(2) There are 119 machines of 65 different types in the sheet metal department of which 60 per cent each exist in only one group.

4.4.(g) Supervision

In the gear centre there were originally two foremen responsible to a shop superintendent, plus three chargehands on days and one on nights, responsible to the foremen. The foremen were functional specialists, one supervising the turning in all groups, the other looking after the remaining processes in all groups. The number of foremen has now been reduced to one. There are no group leaders.

There are four foremen in the sheet metal department for 16 groups.

Group No.	Type of component	No. men	No. m/cs	No. parts	Batch quantity Min.	Max.	Av.	Ops/part Min.	Max.	Av.
A. MACHINE SHOP (Gear Centre)										
1	Timing pulleys	5+2	8	110				4	8	6
2	Small batches gears under 12in dia.	11+5		2500	1	30	15	4	8	6
3	Bevel and worm gears	5+3	12	327				4	8	6
4	Large batches gears under 12in dia.	16+11	35	2500	30	1000	400	4	8	6
5	Large gears over 12in dia.		10	48				4	8	6
6	Speed frame gears	8+5	19	3			250	4	8	6
8	Tufnol gears		3	21	1	30	15	4	8	6
9	Finishing ops	8+1	17	411				2	4	3
		80	104							
	AV:	13	15							
B. SHEET METAL										
1	(B.A.O.) Guillotined blanks	8+8	6	11 000			500	1	2	1
2	(B.A.O.) NC turret press parts	2+3	2	500	25	100				1
3	(A) Ring rails and roller beams	7+2	9	20			350			8
4	(B.A.O.) power press parts	11+2	23	2000	100	5000	1000	1	3	2
5	(O) S/M fabrications	18+2	6		3	50	8	4	20	7
6	(B) S/M parts for 8	18+4	18	6000	1	100	10	1	16	8
7	(A) Cabinets	29+4	16	30	1	20	10	10	20	16
8	(B.A.O.) skilled fabrications	26	9	5000	1	20	10	10	20	16
9	(B) Buckets		5	100			500			7
10	(B) Header tubes grate angles	16	8		32	112				4
11	(B) Spot weld. sub-assy.		4		1	50	20	1	3	2
12	(A) repca. assy.	3	1	1					4	
13	(B) Grate strips		2	3	2000	5000				4
14	(B.A.O.) finishing	3+1	8							2
15	(B) Painting	1	2							
16	(A.O) Parks cramer	17+5	13		1	100				
	AV:	15								

Fig. 4.5 Composition of Platt-Saco groups
Note: In the third column of 'A', 5+2 (for example) indicates 5 men on days and two on nights. In the second column of 'B', (A)=Accrington, (B)=Bolton, (O)=Oldham

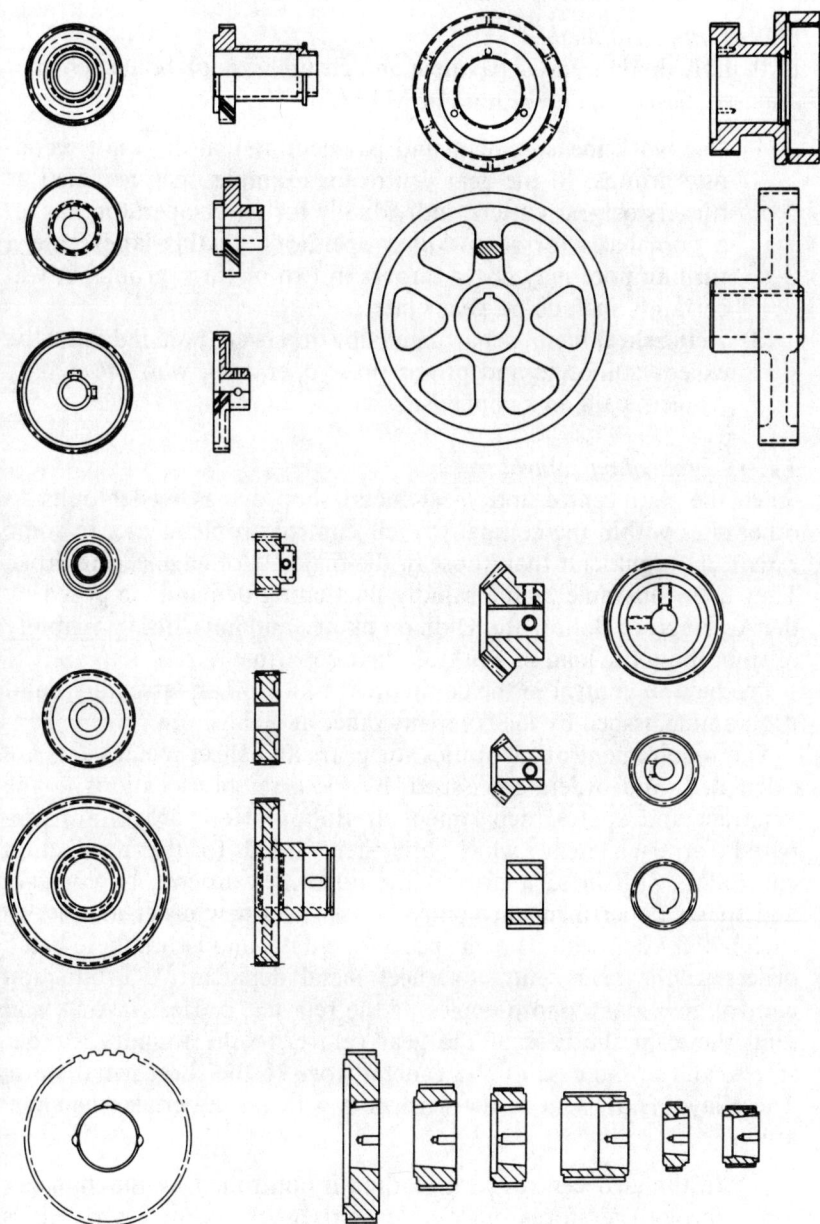

Fig. 4.6 Typical components, Platt-Saco

4.4.(h) Payment method
Both individual incentive payment and group or 'pool' bonus schemes
are used, based on times found by MTM.

(1) The work measurement and payment system does not recognise groups. In the gear centre for example, men are paid as
turners or gear cutters, individually for setter-operators, or on
a pool basis for semi-skilled operators. In this latter case a
turning pool may cover turners in two or three groups, it will
certainly include no gear cutters.

(2) In the sheet metal department all workers are paid individually,
except guillotine and power press operators, who are mainly
women, paid on a pool basis.

4.4.(i) Production control
Since the gear centre and sheet metal shop act as feeder units to
other sites within the company, their control problems are, to some
extent, more difficult than those in the majority of engineering firms.
They are vulnerable to the rapidly fluctuating demands imposed by
the Accrington, Bolton and Oldham plants, and have little possibility
of smoothing the load of work on their departments.

Production control in the company is based on an 'assembly build
programme' issued by the company office at Helmshore.

The requirement of quantities for gears and sheet metal parts are
calculated, and orders are issued, by the assembly locations to the
'contract and spares' department at Bolton. Here orders are prepared for each item, which are then issued to the production
controller, who in turn divides the quantities ordered by contract
and spares department into batches, assigns a new order number to
each batch, each with its own special due-date, and issues these batch
orders to the gear centre or sheet metal department. Production
control also make photo-copies of the relevant process layouts and
send these, in the case of the gear centre, to the foundry, or bar
stores, and in the case of sheet metal work to the sheet metal store.
These layouts are sent to the workshop with the materials when they
are issued.

(1) In the gear centre work loading is controlled by the manager
through a 'stores office'. He holds weekly progress meetings

with the foremen to consider shortages and priorities. This system is working well and at present there are very few overdue orders.

(2) In the sheet metal department a similar method is used, but loading is mainly concerned with Group 1, guillotines, which supplies most of the other groups with cut materials. After this point separate progress men for Accrington, Bolton and Oldham works, arrange for the movement of materials between groups.

4.4.(j) Parts mobility

In the gear centre parts are frequently moved from one group to another for individual operations. This is partly due to the design of the groups – some groups do not have all the machines they need to complete all the parts in their family – but mainly due to efforts to keep machines fully loaded.

In one typical example, of the effect of favouring 'machine loading' over cell discipline, noted by the researchers, two batches requiring first operations on a Herbert No. 4 lathe were issued by the raw material store in succession. The first batch scheduled on the Herbert No. 4 in Group 3 was transferred to the Herbert No. 4 in Group 2, which was idle. The second batch scheduled on the Herbert No. 4 in Group 2, was transferred to Group 3, where by this time the Herbert No. 4 was also idle. For the sake of one hour, cell discipline was sacrificed to 'machine loading' necessitating a switch of both batches back to their proper groups after first operation.

Note. A recent visit to the factory found that this problem has now been solved, and parts mobility has been greatly reduced.

In the sheet metal department there is very high parts mobility. Very few parts are completed in one group. This is almost entirely due to the way in which the groups were designed.

4.4.(k) Labour mobility and flexibility

In both the gear centre and the sheet metal department, there is high labour mobility between groups, for both skilled and semi-skilled workers. This is due partly to wide and unpredictable variations in the demand for different types of component, and partly due to the low flexibility of the skilled setters.

Most setters specialise in setting particular types of machine, so the flexibility of skilled workers is low. On the other hand labour flexibility is fairly high for semi-skilled workers.

4.4.(l) Delegation to groups
Apart from the fact that a few groups do their own setting, there is no delegation to the groups. Because supervision is based on process specialisation, and there are no group leaders, delegation is almost impossible.

4.4.(m) Costing
A conventional total costing system is used based on operation costs and overhead allocation. There has been no change in the costing system to suit Group Technology.

4.4.(n) Design
The change to Group Technology has had no effect on the design function, except that grouping has facilitated variety reduction and standardisation of gears.

4.4.(o) Production planning
A separate production planning office was originally assigned to the gear centre. With a reduction in factory personnel, production planning was re-centralised. With the exception that gear production is planned so that gears are completed in single groups, production planning has not been greatly affected by the introduction of Group Technology.

4.4.(p) Benefits achieved
In the gear centre the benefits obtained were assessed in December 1973 less than a year after implementation. The company claimed:

(1) throughput time reduced from 11 to two weeks;
(2) work in progress reduced from seven to two weeks output, releasing £50 000;
(3) scrap reduced from three per cent in 1970 to 0·25 per cent in 1973, saving £19 000 pa;
(4) output per employee increased 75 per cent giving a direct labour saving of £51 000 pa;

(5) production increased from 2500 gears per week in January 1973 to 8500 per week in December 1973;
(6) quality improved;
(7) variety reduction and standardisation of gear designs.

In the sheet metal shop there has been no detailed study of benefits. It is doubtful if any benefits have been achieved.

The increase in production was due to the consolidation of gear manufacturing for the whole company at Bolton. The other benefits are attributed mainly to Group Technology.

During the last two years the demand for gears has fallen due to a recession in the textile machinery industry. Management state however, that the same benefits are still being obtained although the financial gains have naturally fallen.

4.4.(q) Method of introduction

The introduction of Group Technology was directed by the Manufacturing Director. Outside assistance was provided by consultants from Group Technology International Ltd (GTI).

In the gear centre, the groups were found by component flow analysis, (CFA) using the computer. This work was done by GTI. The initial division into groups in the machine shop found eleven groups. This arrangement was later modified by the company who combined the original Groups 6 and 7 to form a new Group 6, and Groups 10 and 11 to form Group 9. They also in effect combined Groups 6 and 8 which now share the same manpower. This consolidation was carried out partly to facilitate the payment system.

The 'finishing group – Group 9 – does final operations on most of the gears from all other groups. These include grinding, drilling, marking, broaching and metal treatment.

In the sheet metal shop the groups were again initially planned by GTI. The company rejected this layout and made a new layout based on:

(1) groups which specialise in work for the Bolton, Accrington or Oldham factories;
(2) the centralisation of certain processes, e.g. guillotines, Group 1; Wiedeman presses, Group 2 and power presses, Group 4.

As a result, this department has a very complicated material flow system in which most of the groups are heavily dependent on several other groups.

With the possible exception of Groups 3, 7, 8, 9, 12, 13, and 16 which might be called partial groups, this department does not have Group Technology in any normally accepted meaning of the term.

4.4.(r) Date of introduction
In the gear centre re-arrangement of the machines was started in June 1972 and completed in August 1973.

In the sheet metal department the changes were made in 1974.

4.4.(s) Changes in layout
In the gear centre the division into groups has proved satisfactory. There has been no major changes in plant layout, as the consolidation of groups affected adjoining groups.

In the sheet metal shop there have been a large number of layout changes.

4.4.(t) Conclusions
The gear centre is an efficient and well run workshop. It differs from most other Group Technology applications, in that no attempt has been made to achieve a feeling of group identity. This is low partly due to the methods used for supervision and payment and partly due to high labour mobility. Nevertheless, the gear centre, has achieved major economic benefits, probably due mainly to the simplification of the material flow system.

The sheet metal department does not have Group Technology in any usual sense of the term. The organisation into groups, based on the factory supplied, repeats the previous organisation when each factory had its own sheet metal workshop, with the added disadvantages of complicating communications and production control and increasing transport costs. It also eliminates the advantages which might have been obtained by introducing Group Technology. The management are now considering a complete change in layout.

4.5. WHITTAKER HALL LTD, RADCLIFFE, MANCHESTER

4.5.(a) The factory
This company has approximately 90 employees of whom 30 are in the

machine shop. Most are skilled men and these include six group leaders.

4.5.(b) The products

The company makes a range of standard industrial air compressors and vacuum pumps and a range of industrial friction clutches. The parts fall in the medium precision range. Each compressor is assembled with up to 300 various components.

Most compressors and clutches are sold direct to the customer from the factory. However the company exports about half the output of compressors and these are sent to distributors in various countries.

4.5.(c) The production processes

The main production processes carried out by the company are machining and assembly. Out of the 300 parts per product, 50 are wholly machined by the company and a few of the remaining 250 are bought parts which require some further machining in the factory. Fully machined parts are manufactured from either bar stock or castings and the company machines parts from a large range of materials including cast iron, steel, brass, aluminium. All the products are assembled in the factory.

The machine shop contains a variety of machine tools. Since most components are castings, the main operations are turning and boring on surface and boring lathes or vertical or horizontal boring machines. These, and the variety of milling machines, drills, etc., are general purpose machines.

4.5.(d) Where Group Technology is used

There are six groups in the machine shop. These cover all the direct labour and all the machine tools in the department. One of these groups is used for apprentice training. There are four groups in the assembly department, each specialising in a particular range of products. Again all the labour and plant in this department are in groups.

4.5.(e) Motives for introducing Group Technology

The company became interested in Group Technology in 1970. The

Group No.	Type of component	No. men	No. m/cs	No. parts	Batch quantity/month			Ops/part			Total No. of Ops
					Min.	Max.	Av.	Min.	Max.	Av.	
1	Rotors and shafts	3	8	30	1	20	10	5	6	5	
2	Turned components Up to 6in dia.	3	4	50	1	50	30	1	3	2	
3	Turned components Dia. greater than 6in up to 16in	5	6	80	1	20	10	1	4	2	
4	Medium/heavy turning and boring	7	8	30	1	20	10	2	5	4	
5	Heavy turning and boring	2	4	10	1	10	5	1	6	3	
6	Training—mainly turned parts	5	8	40	1	10	5	1	3	2	
A1	Assembly: 20 cfm Compressors	2							10		
A2	Assembly: 50-150 cfm compressors	6							30		
A3	Assembly: 200 cfm compressors	2							10		
A4	Assembly: 300-600 cfm compressors	3							5		
A5	Assembly: special compressors, vacuum pumps and clutches	4							20		

Fig. 4.7 Composition of Whittaker-Hall groups

biggest problems were the unreasonably high stock of finished parts, work in progress and raw castings. Added to this, the sales forecasts proved that there was a limited capacity which might only be increased by further investment in machine tools. The difficulties of improving production control in these situations also caused considerable frustration.

The objectives were therefore:

(1) to increase capacity as quickly and as economically as possible;
(2) to simplify production control and reduce overdue deliveries;
(3) to reduce stocks and work in progress.

4.5.(f) Constitution of groups
The numbers of men, machines and components in each group and the types of components in each family are shown in Fig. 4.7. Drawings of typical components made in the groups are given in Fig. 4.8.

There are 38 different sizes and types of machine installed in the machine shop. Eighty seven per cent of these types exist in one group only, the remainder exist in more than one group.

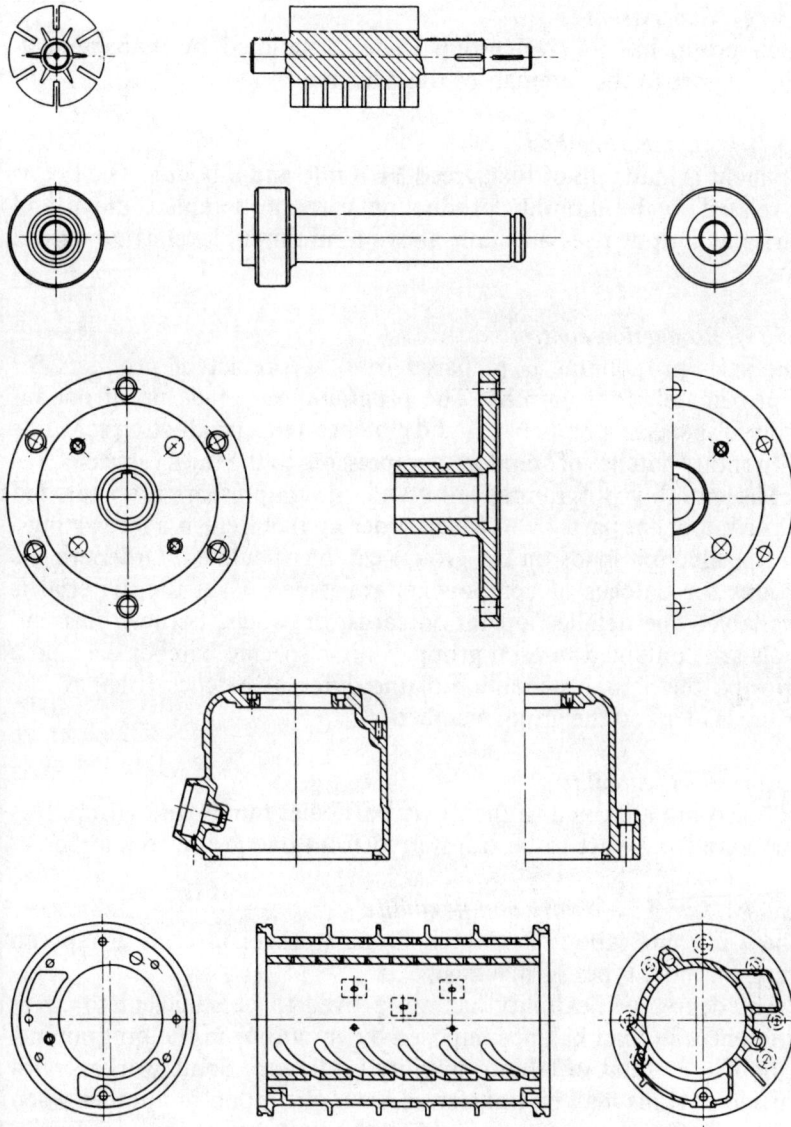

Fig. 4.8 Typical components, Whittaker-Hall
(*Courtesy of IMI Fluidair Compressors Ltd.*)

4.5.(g) Supervision
Each group has its own group leader appointed by management.
They report to the foreman of the machine shop.

4.5.(h) Payment method
Payment is made up of the agreed EEF rate and a bonus. The bonus
is related to the monthly production value in complete units, and
varies as output rises and falls above a minimum level at an agreed
rate.

4.5.(i) Production control
The sales programme is prepared from a forecast of products for
approximately four months. The programme is made up of one-off
orders for special compressors and clutches and a production schedule
of monthly batches of standard compressors to the sales forecast.

The monthly programmes of standard compressors are exploded
into component parts from which ordering requirements for castings,
and production loads on the groups can be calculated. Orders to the
groups for batches of components are issued when the material is
available. The detailed operation cards, drawings, fixtures, jigs and
tools are contained in each group. Unless specific batches become a
priority, the actual scheduling of the different batches is left to the
group leader and the group members.

4.5.(j) Parts mobility
All parts are allocated to their own particular family and group. It is
very rare for a part to be transferred from one group to another.

4.5.(k) Labour mobility and flexibility
There is some labour flexibility. Some workers in each group can
work on most types of machine.

The degree of flexibility has not proved to be sufficient to cover
differences in load balance imposed by variations in the programme.
A limited amount of labour mobility is allowed. Some workers with
particular skills may be transferred from one group to another when
necessary. These transfers are arranged between the group leaders.

There is a small night shift with approximately 15 per cent of the
labour force of the day shift. In this case there is complete mobility

and flexibility. The night shift operators can operate any machine in any group as necessary to overcome bottle necks.

4.5.(l) Delegation to groups

The following activities are delegated to the groups, nominally to the group leader:

(1) setting up;
(2) all inspection;
(3) load scheduling;
(4) distribution of tasks to workers in the group;
(5) tool storage and the storage of production documentation and drawings.

4.5.(m) Costing

A conventional total costing system is used based on operation costs and overhead allocations. There has been little change in the costing system since the introduction of Group Technology.

4.5.(n) Design

The implementation of Group Technology proved a useful opportunity to review the designs of certain components, especially those most recently added to the range. Parts were evaluated according to machining requirements, notably chuck holding.

4.5.(o) Production planning

When introducing new parts, the first task is to allocate them to the appropriate group. Any necessary problems arising in tooling, etc, are discussed with a view to not adding any further tooling when a small design change could suffice.

4.5.(p) Benefits achieved

The company has achieved its main objectives of increasing capacity, simplifying control and reducing late deliveries, and reducing stock and work in progress.

(1) *Capacity.* Unit output from 1971 increased by 50 per cent with a reduction of 10 per cent in direct labour. In addition the new layout provided 20 per cent extra flow space in which further

machines could be added at a later date.

(2) *Deliveries and control.* With a reduction of 50 per cent in the value of work in progress and in the lead time for machining reduced to two months for a production programme, production control and sales forecasting was made easier. This resulted in a quicker throughput of work and a better delivery record.

(3) *Stock and work in progress.* This has been maintained at 1971 value in spite of a doubling of sales output. Turnover is now four times per annum.

	1970	1975
Sales value (£)	420 000	850 000
Stocks raw materials (£)	15 000	12 000
Work in progress (£)	33 000	28 000
Finished component (£)	132 000	160 000
Stock turnover (annum)	2·1	4·1

4.5.(q) Method of introduction

The introduction of Group Technology was directed by the Production Co-ordinator working on an MSc research project and assisted by a team from UMIST.

The groups were found by classifying and coding the component drawings, totalling approximately 1500. A major aim in forming the groups was that all the parts should be completed within their own group.

4.5.(r) Date of introduction

The groups were formed in 1971 after eighteen months study. The machines in the company were all physically moved during a two week holiday shut-down period.

4.5.(s) Changes in layout

The division into groups has proved to be stable. No change in the group layout has been required since introduction.

4.5.(t) Conclusions

This is a successful application of Group Technology. It is important because it shows that even in relatively small companies Group Technology can be introduced and can give major advantages.

CHAPTER 5

BENEFITS OF GROUP
TECHNOLOGY

5.1. INTRODUCTION

The two previous chapters described the changes made by a number
of British companies from traditional forms of organisation to
Group Technology. This chapter analyses the benefits which have
been obtained.

The first part of the chapter is based on studies made by research
teams at Salford and Bradford Universities and at the London
Business School into the operational and economic benefits of Group
Technology. The second half of the chapter looks at the social bene-
fits and is based on studies made by a research team at Birmingham
University.

As far as the economic benefits are concerned, the research has
provided evidence that the change to Group Technology in machine
shops can lead to major economic benefits. Such changes as a reduction in work in progress by 62 per cent, a reduction in throughput
time of 70 per cent and an increase in output per employee of 33 per
cent are only the average from ten of the best companies.

There is also evidence that the change to Group Technology does
provide a 'climate' where important improvements in job satisfaction, motivation and industrial relations are possible, which cannot
be achieved with our traditional forms of organisation. It is difficult
and perhaps impossible to obtain objective measures of social variables
such as job satisfaction and motivation. Nevertheless the research has
indicated that Group Technology does lead to changes in a number
of important social factors. There is considerable research evidence
that these types of change tend to increase job satisfaction.

The research has shown that there are few automatic benefits of
Group Technology. What Group Technology does is to provide
conditions where much greater savings can be made than would ever
be possible with traditional methods of organisation, providing that
management takes the necessary action to achieve them.

103

5.2. OPERATIONAL BENEFITS

Before considering the economic and social benefits of the change to Group Technology, it will be helpful to consider in practical terms what the change means to production management. What in other words are the operational benefits which can be achieved.

5.2.(a) Simplification of the material flow system

The first of these operational benefits is a major simplification of the material flow system. The way in which Group Technology simplifies the material flow system is illustrated in Fig. 5.1. Figure 5.1 (a) illustrates the technological system in a particular factory. In other words it shows the relationship between the different processes and machines used to make a company's range of products. In companies organised on traditional lines, in which the organisational units are based on the process, Fig. 5.1 (a) also illustrates the organisational system of the company. In addition, because materials flow between these organisational units, Fig. 5.1 (a) also illustrates the material flow system in the company.

Figure 5.1 (b) illustrates the fact that the organisational and material flow systems do not necessarily have to mirror the technological system. With only minor changes in the technological system it is possible to re-organise into major groups or departments, each of which completes a major family or set of products at a particular major stage of processing. It will be noted that these major groups have the same names as many of the departments found in factories today. The only difference between these traditional 'departments' and 'major groups' is that the major groups have been organised so that they can complete all the operations on all the parts they make. These major groups fit the definition of a group given in Chapter 2, with the sole exception that they usually employ many people. It has already been noted in Chapter 2 that this change greatly simplifies the control of material flow between departments.

Figure 5.1 (c) illustrates the fact that these major groups or modified departments can be divided into smaller groups, each of which completes a particular family of components. This change further simplifies the flow of materials through the factory.

The simplification of the organisational and material flow systems gives a number of major advantages to production management.

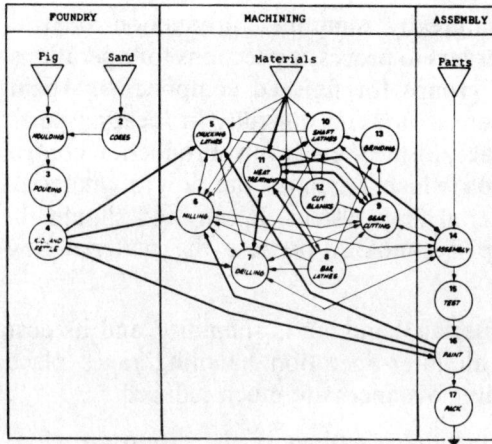

(a) *Technological systems network*

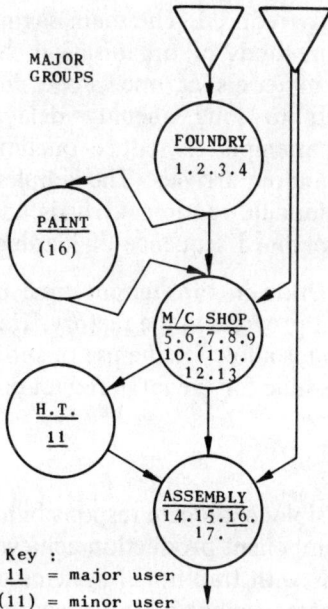

Key :
<u>11</u> = major user
(11) = minor user

(b) *Major groups*   (c) *Groups*

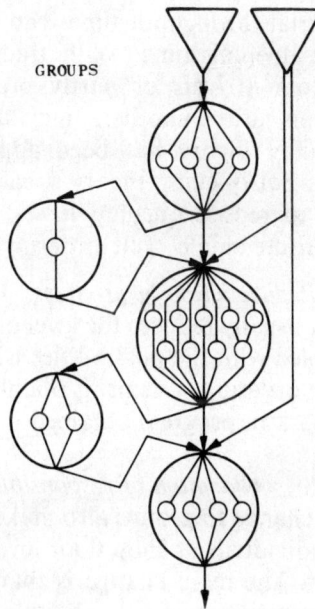

Fig. 5.1 Functional to group organisation

(This illustration is an extract from Final Report on a Study into the Effects of Group Production Methods on the Humanisation of Work, *published by the ILO's International Centre for Advanced Technical and Vocational Training, Turin, Italy, and is presented here with their approval as copyright holders)*

(1) *Production control.* It greatly simplifies production control, because instead of issuing orders to processing sections for operations, they now issue orders to groups for finished components. Again, progressing is simpler. There is now no difficulty in locating parts. They can only be in one place, inside their group. Production control is the management function which regulates the flow of materials through a company's material flow system. It can, and should, be much simpler with Group Technology because the material flow system is simpler.

(2) *Materials handling.* Materials handling is simplified and its cost can be reduced, because all inter-operation handling takes place inside the group and handling distances are much reduced.

(3) *Throughput time.* Because the machines in the group are close together and because they are all controlled by the same foreman, material throughput times can be greatly reduced. The main saving is in queuing time. With traditional methods of organisation the random arrivals of partly processed materials at one processing section after another, inevitably lead to long queuing delays. Queuing theory has been applied in attempts to reduce queuing time, but queuing theory is based on random arrivals. The simplest way to reduce queuing losses is to eliminate random arrivals and substitute single cycle ordering and a planned sequence of loading.

(4) *Reduce work in progress.* The reduction in throughput time of itself usually reduces the level of work in progress in the factory. This coupled with the possibilities which groups allow for the use of short cycle ordering systems, is mainly responsible for the large reductions in work in progress obtained in practice.

5.2.(b) Allocation of responsibility

The change to groups also makes it possible to allocate responsibility to individuals at shop floor level for component production achievement. The main change is that whereas with traditional systems of organisation it is only possible to allocate responsibility to sections for the completion of operations, with Group Technology it is possible to allocate responsibility to groups for the completion of components.

With traditional systems of organisation it is essential to have detailed central planning and control in order to co-ordinate the work of the different process specialising sections which contribute to the manufacture of complete parts. With groups the responsibility for quality and for delivery by due-date can be effectively allocated directly to the foremen of the groups, because they control all the means necessary to achieve the required result.

Many companies which have introduced Group Technology have allocated the responsibility for quality to their groups. Because the foreman and workers in a group see the effect of each operation on the following operations, they are in a better position to ensure that the finished components conform with the required quality standards than are their colleagues in traditionally organised 'sections', who only see the results of one process. Again the foreman of a group who receives at the beginning of each period a list of all the components to be completed by the end of the period, has all the information needed to plan the schedule of work in order to achieve this target.

5.2.(c) Reduced setting time and increased capacity
Providing that the foreman of a group receives a batch of orders at the beginning of each period, he can plan the sequence of loading the parts on his heavily loaded machines, in order to bring together parts with similar set-ups and obtain major reductions in setting time. Some saving in setting time can be achieved with these methods even with traditional forms of organisation. Much greater savings are possible, however, with Group Technology. Reducing the time during which machines are being set up, increases the time during which they can do useful work. Reducing setting time therefore increases the capacity of the group and of the factory to produce finished products.

This method of reducing setting time can be applied to any machine with two or more tool setting positions. The potential savings are greatest on multi-tool machines such as capstan lathes and automatics. The potential increase in capacity is therefore greatest where these types of machine are the most heavily loaded. This benefit was the first benefit claimed for Group Technology. The present research has indicated that potential savings in setting time have been ignored in many of the more recent applications.

5.2.(d) Reduced data processing
The simplification of the material flow system greatly reduces the need for paper work in the factory. In the case of factories making standard machines for example, the permanent data required for production such as route cards and drawings, can be stored in the groups. The orders on the other hand are now concerned with components to be made and not with the much greater number of operations to be carried out. In several companies using Group Technology today the only order paper work issued to the groups is a list of components to be finished, which is issued to each group at the beginning of every period, together with a load summary showing the load in machine hours on each machine in the group, imposed by this list order.

5.2.(e) Improved commercial liaison
Particularly where the production control system has changed to single short cycle ordering, production can now provide a flexible system which can change quickly to meet changes in market demand. It is possible under these conditions for production to have a series of 'firm' production programmes for only short periods ahead, and for marketing to be able to change their sales forecasts at these regular short intervals and thus to follow closely changes in market demand, and product mix.

Probably the main advantages of Group Technology to marketing are that the shorter throughput times make it possible to quote shorter deliveries, and that the simplified material flow system and unified responsibility for products, ensure more reliable quality and delivery.

5.2.(f) Production design liaison
With Group Technology it is simpler to introduce new products, or modifications to existing products. Each component affected is made in only one group. The problems of introducing a new component into production or modifying an old one, are tackled together inside one organisational unit.

5.2.(g) Production technology liaison
The change to Group Technology also simplifies the introduction of

(1)	(2)	(3)	(4)	(5)	(6)	(7)	(8)	(9)
	Net despatches	Value added	Average no. employed	Total wages and salaries	Despatches/ employee	Output (tons)/ employee	Value added/ employee	Average income/ employee
	£M	£M		£M	£		£	£
1961/62	2·220	1·615	1001	0·714	2218	4·0	1613	714
1962/63	2·184	1·580	952	0·706	2294	4·5	1660	742
1963/64	2·585	1·872	903	0·752	2863	5·5	2073	833
1964/65	2·922	2·007	941	0·844	3105	5·8	2133	897
1965/66	3·363	2·303	992	0·951	3390	6·0	2320	953
1966/67	3·768	2·646	987	0·979	3818	6·5	2681	992
1967/68	3·576	2·519	906	0·951	3947	6·9	2780	1050
1968/69	4·984	3·117	1130	1·338	4411	7·4	2759	1184
1969/70	6·008	3·786	1181	1·578	5087	7·5	3206	1336
1970/71	6·727	4·374	1179	1·787	5706	7·9	3710	1516

Fig. 5.2 Economic effects of Group Technology at Serck Audco
Note: During 1968/69 the progress in column 8 was retarded by the problems associated with the transfer of £1.3m of ball valve production to Newport

(*Reproduced from* Production Planning and Control, (*HMSO*))

new methods and new machines into production. As these changes are generally concerned with particular parts or types of parts, they are concerned also with particular groups. The production technologist does not have to deal, therefore, with a number of different foremen who are all affected by the change, he only has to deal with one group.

5.3. ECONOMIC BENEFITS

The operational benefits just described represent an important simplification of the work of the production manager. They make it possible for him to regain control of the system. These operational benefits have also led to major economic benefits in a number of companies.

5.3.(a) Serck Audco Ltd

Many companies are unwilling to publish the economic benefits they have achieved by changing to Group Technology. The only comprehensive table of results worthy of the name comes from the pioneers, Serck Audco Ltd, and is shown in Fig. 5.2. The results in columns 6, 8 and 9 contain an element of inflation, but there is no denying the improvement in column 7 where the output per employee is expressed in real terms, i.e. in tons of valves. This clearly shows that the initial spurt of a 50 per cent improvement over the four years up to 1965/66, has not levelled off but has been maintained and improved as the operating methods within the groups have been improved.

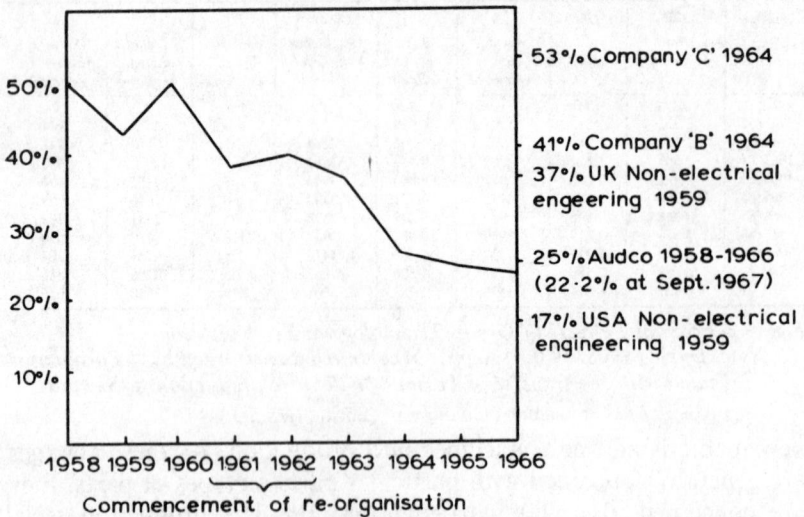

53°/₀ Company 'C' 1964

41°/₀ Company 'B' 1964
37°/₀ UK Non-electrical engeering 1959

25°/₀ Audco 1958-1966 (22·2°/₀ at Sept.1967)

17°/₀ USA Non-electrical engineering 1959

Commencement of re-organisation

Fig. 5.3 Serck Audco's stock/sales ratio
Note: Audco and companies B and C are in the same sector of industry, are partly competitive and are similar regarding product varieties and batch sizes; each employs around 1000 people
(*Reproduced from* Production Planning and Control, (*HMSO*))

Additional information about Group Technology at Serck Audco, (20) previously published by the EDC and relating only to the period from 1961/62 to 1966/67, is summarised in the table below.

1. Sales	Up 32 per cent
2. Stocks	Down 44 per cent
3. Stocks/sales ratio	Down from 52 per cent to 22 per cent
4. Manufacturing time	Down from 12 weeks to 4 weeks
5. Overdue orders	Down from 6 weeks to under 1 week
6. Output per employee	Up about 50 per cent
7. Capital investment	Cost recovered four times by stock reduction alone.

The reduction in stocks of 44 per cent included a reduction in work in progress of 67 per cent. The improvement in Serck Audco's stocks/sales ratio over eight years is shown in Fig. 5.3. It will be seen that the company changed its rate of stock turnover from twice a year in 1958 to nearly five times a year in 1966. The two other companies (b) and (c), shown in this illustration for comparison, are both from the same sector of industry, are partly competitive and are similar regarding product varieties and batch sizes. All three companies employ around 1000 people.

5.3.(b) Benefits published by other companies
These results of Serck Audco have been so widely quoted that there would be no point in repeating them if they could not now be compared with the results achieved in other factories. Research carried out at the London Business School into the published claims of other companies which have changed to Group Technology, shows that at least ten other companies have achieved or beaten the remarkable results claimed by Serck Audco Ltd (23)–(31).

	Serck %	*Max.* %	*Average* %
Reduction in work in progress	67	85	62
Reduction in stocks overall	44	44	42
Reduction in throughput time	66	97	70
Increase in output/employee	50	50	33
Increase in sales	32	—	—
Reduction in overdue orders	85	—	82

Fig. 5.4 A comparison of benefits achieved by Serck Audco and nine other companies

Figure 5.4 shows six of the economic measures which were compared and gives the results claimed by Serck Audco and the averages achieved when the other companies which have published their results, are included. Each of these measures will now be considered in turn.

(1) *Reduced work in progress.* By far the most widely quoted benefits of Group Technology are a reduction in work in progress and in throughput time. These two are obviously very closely related.

Ten companies claim reductions in work in progress ranging from a 45 per cent reduction by Thomas Mercer Ltd to an 85 per cent reduction claimed by Ferodo Ltd. The majority of companies were close to the average reduction of 62 per cent.

The reduction claimed by Serck Audco Ltd of 67 per cent is therefore slightly above average. In fact the real reduction achieved by Serck Audco is higher than 67 per cent because this reduction was achieved in association with a 32 per cent increase in sales. Thus for a constant sales figure the reduction achieved was actually 76 per cent. A reduction of this magnitude was achieved by only two other companies in the sample.

(2) *Reduced throughput time.* The second most quoted benefit of Group Technology is a reduction in throughput time. The 66 per cent claimed by Serck Audco seems to be on the modest side compared with claims reaching as high as 97 per cent and an average of 70 per cent.

The overall effect of reduced throughput time will depend on other factors such as the type of product. In an organisation whose final products are assemblies, the reduction in throughput time of components will only cause a reduction of the investment in parts held in the factory between material supply and assembly. The effect on customer service may be limited to that caused by a reduction in the number of shortages.

In companies whose products are components or who are in the position of making only to customer's orders, the reduction in throughput time is reflected directly in better customer service.

(3) *Output per employee.* Serck Audco claimed an increase of 50 per cent in output per employee through the introduction of Group Technology. One other company, the Langton Company of New Jersey, claimed a similar increase in productivity over the period spanning the first half of 1969 to the first of 1970. Including the Langton figure, the average published productivity gain is only 33 per cent and without it, the average falls to 28 per cent.

Many companies are reluctant to reduce their work force knowing that skilled labour is difficult to recruit and that redundancy has a very disturbing effect upon the rest of the work force. The actual improvements in productivity achieved with Group Technology, may well be masked by a reluctance to reduce the labour force while there is still hope that sales will improve.

From experience in visiting companies using groups, it is believed that these considerations are often dominant in the minds of management. Thus it is believed that the published average of a 33 per cent increase in output per employee is slightly low and that the 50 per cent achieved by Serck Audco is quite possible when a system has settled down due to natural wastage of the work force.

(4) *Increased sales.* Serck Audco is the only company in the sample to claim an increase in sales. There are obviously factors which might have contributed to this increase other than Group Technology. It was included by Serck Audco because they believed that the increased capacity and better deliveries due to Group Technology, were responsible for the increase.

(5) *Reduction in overdue orders.* Only two companies claimed a reduction in overdue orders. They were Serck Audco and GEC Elliot. Both of these companies have made substantial moves towards Group Technology. It may be that only a company wide approach to Group Technology will have a measurable impact on customer service. Both companies claimed substantial reductions in overdue orders, the average of the two being 82 per cent.

(6) *Reduction in total stock level.* Again only Serck Audco and GEC Elliot claimed reductions in total stock level. These were 44 per cent and 40 per cent respectively.

This study shows that not only are the economic claims made by Serck Audco consistant with those made elsewhere, but that in some cases they have been exceeded. In general the claims made by Serck Audco are not the highest achieved and they should, therefore, be achievable in other similar applications.

The study described above was restricted to only six economic benefits. There is considerable published evidence that Group Technology gives other economic benefits as well. For example, several companies have reported major reductions in their tooling investment, due to the use of common tooling for 'tooling families' of similar parts.

Big reductions have been reported in setting costs, due to reductions in setting time, and in handling costs, due to the reduction in handling distances when all inter-operation handling moves take place inside groups.

Finally a number of companies have reported reductions in indirect labour, due mainly to the simplification of control with a simpler material flow system. This reduction was partly responsible for the major increases in output per employee in the companies studied by the London Business School.

It is submitted that economic benefits of the order reported here, could not have been obtained in any of these companies with their original traditional forms of organisation.

5.4. ECONOMIC BENEFITS IN ASSEMBLY

The economic benefits reported so far, were obtained mainly in machine shops and other component processing applications. The economic benefits achieved by changing to groups in assembly have not been generally of the same high order. Most of this experience has been outside the UK. It is reported here to emphasise the fact that it is in component processing where the main economic gains can be expected from the introduction of Group Technology.

Many of the assembly groups found in Europe were formed to replace mass production assembly lines. What ever the failings of these assembly lines, they do normally have short material through-put times and a low investment in work in progress. The possible economic savings are therefore relatively small.

The reasons why groups have been introduced in this case, are mainly social. The increase in demand in the 1960's led to an increase in the number of work stations on many assembly lines, thus reducing the work content and cycle times at the stations. This made jobs on the line unattractive to many european workers. The short term solution was to import foreign workers from countries with a lower standard of living. This, however, brought its own problems. As one director of a european car company stated at the time, 'There is no long term future in the importation of second class citizens to do the jobs which our own nationals won't accept. Our aim in changing assembly methods is to make the work attractive to our own people'.

When group methods were first planned for assembly, it was generally expected that costs would rise. It was reasoned that if people couldn't be persuaded to work on lines, they ceased to be an economic proposition. Change was therefore necessary, even if it increased costs. In the event, most companies have claimed some economic benefits from the change. For example the cost of assembling the Volvo car at the new Kalmar plant, is estimated to be 10 per cent less than it would have been by conventional methods. Philips too have claimed that the change to groups increased the output of black and white television sets per employee by 12 per cent. These benefits appear to be partly due to a reduction in line balancing losses and partly due to increased motivation.

It is interesting to note that a small line, employing for example fifteen workers, fits the definition of a group and can have all the desirable characteristics of a group described in Chapter 2. Many of the assembly groups found in practice, are in fact organised internally as small lines. It seems that lines only become objectionable to workers when they are very long and the jobs on them are fragmented, have a very short cycle time and are machine paced.

5.5. SOCIAL BENEFITS

The researches covered in the early part of this chapter were mainly concerned with operational and economic benefits. The observations made during the research, did, however, support the findings of other research projects, which indicate that workers generally prefer the work in factories organised for Group Technology, to work in traditionally organised factories.

5.5.(a) Studies based on questionnaires and interviews

One of the difficulties in measuring social benefits, such as increased job satisfaction, is that there are no known objective methods for measuring the opinions or feelings of human beings. As was demonstrated by the Hawthorne Experiment (21), experiments with human beings tend to change their attitudes and behaviour. The final step in a controlled scientific experiment, of a return to the original conditions, is probably impossible. Judgements about social benefits must therefore be mainly subjective.

At first sight the obvious way to measure job satisfaction is to ask the workers affected by a change, what they feel about it, either by getting them to fill in a questionnaire or by interviewing. Most of the research into the effects of group production methods on job satisfaction, has used this approach, and their results have been over-whelmingly positive in favour of groups.

It was to be admitted, however, that these methods are not objective. Research has shown that the answers given can vary widely due to outside influence which have nothing to do with conditions at the work place. They can vary with minor changes in the drafting of the questions; with the time at which they are issued, and the answers will depend on who has sponsored the research and issued the questionnaire. With interviews again, the results achieved will depend greatly on the behaviour and attitudes of the interviewer and will also depend on his interpretation of what he hears. It can be said, however, that the fact that these researches have all given positive indications of increased job satisfaction with the change to groups, does provide strong subjective evidence that this is the normal effect.

5.5.(b) The measurement of social factors

Another approach to the promotion of job satisfaction is to design organisation systems and methods of organising work which foster its development. There is a great deal of sociological and psychological research, which indicates that job satisfaction is closely associated with certain key factors. These include, for example:

(1) the degree of association with the completion of significant products;
(2) the possibility of using to the full one's knowledge and skills;
(3) belonging to a group;
(4) association with a territory;
(5) some say in the planning and control of one's own work.

These social factors can be measured objectively. The aim when designing an organisation system should be to achieve results which exhibit high values for these and other related social factors. There is no *guarantee* that making such changes will in all cases increase job satisfaction; there are grounds for hope, however, that they will provide conditions with greater possibilities for improvement than our present systems. The study undertaken by Birmingham University, which will now be reported, studied the effects on these social factors of changing from a traditional system of organisation to Group Technology.

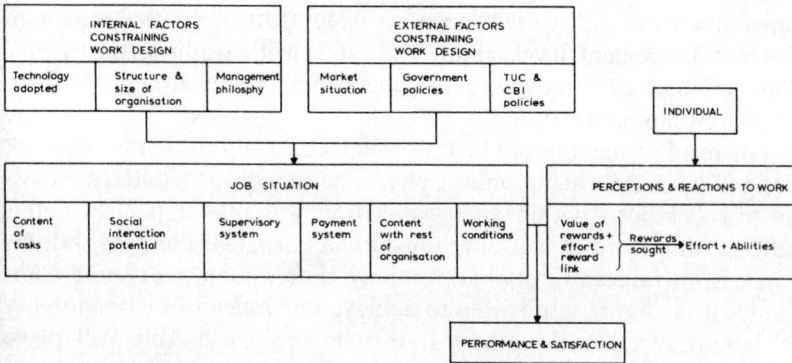

Fig. 5.5 Job features and behavioural attitudes

5.6. GROUP TECHNOLOGY AND JOB SATISFACTION

To evaluate the benefits of Group Technology it is important to establish what changes its introduction will make to the job features that affect behaviour and attitudes. Figure 5.5, illustrates these relationships. It illustrates the idea that the workers performance and job satisfaction are determined by the interaction of the job situation and the workers perception of, and reaction to, his job.

5.6.(a) Internal factors constraining work design

First of all it will be noted in Fig. 5.5 that there are a number of internal factors inside any company which act as constraints on the way in which work can be organised. The first of these is the type of technology adopted. In many cases there will be a range of technologies from which a company can choose when deciding how to produce their goods. Once a decision is reached, however, and a particular technology has been chosen, it places constraints on the choice of job content and work organisation.

A second internal constraint is the structure and size of the organisation. As an organisation expands, top management is forced to decide on the size of its operating units and on the management structure they will adopt. The choice that they will make will also impose constraints on possible methods of work organisation. Group Technology itself has some effect on this choice. With Group Technology it is possible, in some technological systems, to keep operating units small but still to work with a wide span of control and with fewer management levels than is possible with traditional systems. This change can reduce communication problems and lead to improved labour relations.

The third of the internal factors which constrain the choice of work design, is management philosophy. The established pattern of industrial relations within an organisation will play a part in determining what form of work organisation a particular company adopts when it introduces Group Technology. For example, in some companies it is almost impossible to achieve much flexibility or mobility of labour, whereas in others it is relatively easy. This will place constraints on the extent to which job enlargement can be introduced with the change to Group Technology.

5.6.(b) External factors constraining work design

Certain external factors will also provide constraints which limit the possibilities of a choice in work design. The first of these is the labour market situation. If the labour market is tight, and it is difficult to attract and retain workers, the organisation is more likely to introduce Group Technology because of its potential for increasing both job satisfaction and output per employee.

Government policies can also provide constraints which will affect work design. Such policies as nationalisation, pay restraint, price controls, taxation policies and policies controlling the location of industry, will all have some effect on the choice of work design and on the possible effectiveness of any given choice.

Finally the unions and the employers federation will also provide constraints. How favourably the unions and employers federations view the trend towards increased employee participation at shop floor level, for example, will affect whether or not companies use the opportunities provided by Group Technology to increase participation.

5.6.(c) The job situation

The diagram in Fig. 5.5 shows how performance is determined by the interaction of the job situation and the individual's perception of, and reactions to, the job. The job situation can be viewed as comprising six components. These are the content of tasks, the social interaction potential, the supervisory system, the payment system, the working conditions and the contact with the rest of the organisation.

The effort that an individual makes at work depends on his perception of his job situation. This effort will be directed towards gaining the specific rewards he believes he can obtain in the situation. His overall performance is affected by two additional factors; his abilities and his perception of his role. Thus, what is rewarded is of central importance to personal performance in that whatever behaviour is perceived as rewarding, will be pursued to the best of the person's ability in order to obtain that reward.

The worker has to make a decision about which rewards he values most highly. For instance a worker may value the reward of the social esteem of his work mates and also value the higher level of earnings he achieves by producing at a high output level. If his work group has set norms limiting output, he has to choose between the two potentially rewarding actions.

5.6.(d) Job content

The jobs done by workers in industry can vary between those with very low content, such as the jobs on mass production assembly lines, and those with very high content, comprising a great number of different tasks. The rewards that an increase in job content can offer are:

(1) *Self fulfilment*, through using skills and abilities which are valued by the worker. People perform at high levels when they are using the skills and abilities which they have and value.

(2) *Sense of achievement.* This is different from self fulfilment in that it is possible to achieve a target using abilities which are not particularly valued. Satisfaction comes in this case from the achievement of the objective rather than because certain potential qualities were utilised and made achievement possible.

(3) *Feeling that one's contribution is valued.* Praise and respect from managers, supervisors, and fellow workers are valued rewards.

(4) *Stimulation*. People need a certain amount of novelty, complexity and uncertainty in life. At work they may find this stimulation in the variety of tasks, or in the variety in the environment.

(5) *Autonomy*. Some degree of self determination at shop floor level can in some circumstances provide an incentive to performing responsibly.

(6) *Self development* by the acquisition of new knowledge and skills can itself be seen as rewarding.

For the realisation of many of these rewards some feedback is necessary about the effect resulting due to a person's actions. To utilise potential skills effectively or to achieve a particular target, a worker needs a feedback system which gives him information about his output and errors made and enables him to monitor his performance against the target.

The change to Group Technology increases the possibilities for the achievement of many of these rewards. For example, job enlargement has been seen as one method by which the reward of self fulfilment can be increased. In recent years there have been many experiments to test the effects of job enlargement on workers. In some cases this has involved increasing the number of tasks of the same type, that an employee performs. For example he may be trained so that in addition to operating lathes he can also operate other types of machine. This is known as 'horizontal enlargement'. In other cases tasks of a different level are added to the existing job. For example, he can be trained to set-up his own machine and to inspect his own work. This is called 'vertical enlargement'. Research has shown that most but not all workers prefer enlarged jobs. Group Technology can facilitate a move towards enlarged jobs and thus make work more rewarding for employees. For example, the fact that each group contains a number of different types of machine, increases the possibility of horizontal enlargement, provided that the workers are prepared to accept some flexibility.

In addition the change to Group Technology makes it possible to retain efficiency and at the same time to delegate many decisions to the group. In other words it makes it possible to introduce a large measure of vertical enlargement, which would be impractical with traditional methods of organisation.

Group Technology also increases the potential for rewards due to a sense of achievement. Simple and clear targets and objectives can be stated in terms of completed products and time. This makes it easier to measure achievement and to obtain satisfaction when achievement meets the target.

Group Technology will also have some effect on the reward of 'stimulation'. In some cases Group Technology may bring with it a decrease in variety. For a few workers the change may reduce the variety of different operations which they carry out. On the other hand, variety and stimulation may increase if the operators are trained to be flexible, as the differences between the operations they perform and the differences in the machines they work, will be greater than under a traditional system.

A reduction in the variety of personal contacts may be brought about if there is a policy of geographically isolating groups from each other. However, although this may reduce the variety of potential stimuli, it will tend to encourage feelings of group cohesion and belonging which can be of greater importance for the members of the group. In some cases the number of people any one individual contacts may increase with Group Technology. This can be so if the group are responsible for making decisions, which bring them into contact with staff departments. Again the change to Group Technology makes it possible to delegate types of decision making to the floor of the shop, which it would be impossible to delegate with traditional forms of organisation. It does therefore increase the possibility of obtaining the rewards of autonomy or self determination.

Finally, the change to Group Technology makes it possible for the individual worker to feel that his contribution is important. He is now associated in the completion of easily recognised, significant and completed products, and can see where his contribution fits into the production of that product.

5.6.(e) The social situation
The second factor in the job situation illustrated in Fig. 5.5 is the social interaction potential. The possible rewards in this case are

social esteem and affection. The extent to which an individual needs these rewards will determine the extent of his conformancy to the accepted practices of the social group within which he works. Group Technology can result in work groups with a sense of identity and with strong group cohesion. If the group objectives are in line with management objectives the increased group cohesion will have a positive effect on performance. If however the group is not committed to meet management targets, performance may suffer.

5.6.(f) Supervisory style

The third factor in the job situation is the supervisory system. The behaviour that a supervisor rewards and punishes certainly affects performance. If he is perceived as considerate and interested in his subordinates, and if there is mutual trust between them, certain aspects of performance are likely to be better than if he is not valued in these ways. This assumes a conventional structure with control of decision making resting in the hands of the supervisor. As has been mentioned above, research has shown that the ability to control one's environment is a reward which many people value highly. One of the ways in which such control can be increased is by the introduction of shop floor participation.

5.6.(g) The payment system

Here the rewards may include both the absolute amount of money earned and the perceived fairness of the pay received in relation to that received by others with comparable abilities and training.

The change to Group Technology has tended to re-enforce a trend away from individual incentive payment schemes towards group bonus or day work. Group Technology makes it possible to set output targets in units of completed parts which the traditional system in the batch manufacturing industry does not allow. The fact that simple targets can be set, helps to provide achievement motivation and makes it easier to monitor performance. It is possible that this change will improve industrial relations by reducing the conflict over incentive payment rates. Again if Group Technology leads to greater efficiency through shorter throughput times, greater accountability and less work in progress, there is a potential for increased wages.

5.7. CONCLUSIONS

The research described in this book shows that the change to Group Technology can bring with it many major benefits.

First of all it can result in operating benefits due to the greater simplicity of the material flow system, the organisational system, and the communication system. The effect of this simplification of the production system is that it makes it much easier to control. Much of the stress and confusion associated with production under the traditional forms of organisation, is eliminated.

Due to the simplification of the production system, major economic benefits can be achieved. It is now certain and no longer a subject for dispute among intelligent observers, that the change to Group Technology, particularly in the case of component processing workshops, can make possible major economic benefits. Many of these benefits do not arise automatically from the introduction of Group Technology, but require appropriate management action to achieve them. These benefits can, however, be much greater than is possible with traditional methods of organisation.

There is also considerable evidence that the change to Group Technology can lead to improved worker job satisfaction and motivation. While these social factors are very difficult or perhaps impossible to measure objectively, it is fairly easy to demonstrate that the change to groups can provide an environment where increased job satisfaction should be possible. The change can, in other words, eliminate many of the features of the traditional form of organisation, which have been criticized by sociologists, by pyschologists and by the workers themselves.

CHAPTER 6

DESIGNING GROUPS

6.1. INTRODUCTION

The early chapters of this book have shown that changes in industrial organisation are urgently needed and have demonstrated that Group Technology provides a new approach to the organisation of work which can overcome many of our existing problems.

This chapter now tackles the problem of how to set about re-organising a factory into groups. The recommendations given are based on studies done at Bradford and Salford Universities and at the London Business School into the methods used by companies which have already introduced Group Technology.

The design of groups takes place in three main stages. First it is necessary to decide on the policy for introduction, to define objectives, to plan the production system required to meet these objectives, and to make any changes in departmental organisation which are needed before groups can be introduced.

Second comes the identification of the components to be made and of the machines to be installed in each group. Techniques for planning this division into groups are described in the chapter.

Finally it is necessary to plan the manning, management, layout and services for the groups and to plan any necessary changes in supporting systems.

6.2. GENERAL POLICY

When deciding on a policy for the introduction of groups there are three main types of decision to be made. These are concerned with:

(1) preliminary changes;
(2) general approach;
(3) who should undertake the design and implementation exercise.

6.2.(a) Preliminary changes

Two approaches are possible. It is possible to change to groups using the existing machines, production methods, and product designs.

125

Alternatively it is possible to precede the change to groups by extensive preliminary changes in product design and in production methods.

Using the second method some companies have carried out major exercises in variety reduction, and in methods and tooling development, and have purchased and installed new and more advanced types of machine, before attempting to form groups. On balance, however, the first policy which changes to groups without prior changes in design or methods is probably the best. The main reasons for this opinion are as follows. First, in relation to product design, the change to groups will greatly simplify the planning of effective variety reduction. There is a high probability that those parts which can be eliminated in favour of one standard design, will all exist in one group only.

Second, the types of technological development which are desirable with groups are likely to be different to those which appear attractive with traditional forms of organisation. The purchase of some new types of machine may, in fact, hinder the introduction of Group Technology rather than help it.

Third, the change to groups is itself a major innovation. To many it appears unwise to tackle both a major organisational change and a major change in technology both at the same time. Finally the change to Group Technology tends very quickly to release capital previously tied up in work in progress and other stocks. Less new capital will be required therefore for the expensive technological development stages if this is left until after groups have been introduced.

6.2.(b) General approach to design and implementation
For the design of groups and the implementation of Group Technology, there are four main approaches possible.

(1) Comprehensive analysis and comprehensive implementation.
(2) Comprehensive analysis and pilot implementation.
(3) Sample analysis and pilot implementation.
(4) Sample analysis and comprehensive implementation.

Comprehensive analysis is based on a study of all the components produced in a machine shop. Sample analysis on the other hand, is based on only a sample of parts. It is possible to plan a division into groups and families with either method.

Comprehensive implementation requires that a total plan should be prepared for the division into groups and that when this plan has been adopted the implementation of the plan should be continuous. Pilot implementation on the other hand starts by introducing a pilot or trial group and then adds new groups intermittently as the opportunity arises or when there is a demand for new groups from the workers in the factory.

Although all these approaches have been used, there is no doubt that the first has in practice provided the best results. The use of analysis based on a sample of parts may be unavoidable in some cases. For example in jobbing production the parts produced are changing all the time and the only possibility for analysis is to base it on a sample of past orders. In most other situations, while a sample analysis may save some initial expense in comparison with comprehensive analysis, it will probably involve greater expense in the long run due to the need to redesign and reorganise groups when parts not included in the sample become due for production. It is also impossible with a sample to make the accurate load checks needed for a reasonable distribution of common machine types between the groups.

Implementation is discussed in more detail in Chapter 7, but it is worth noting here that there are several limitations associated with pilot implementation. Pilot implementation for example, may mean operating the production system for a long time with different production control and wages systems for the groups and the remainder of the work shop. Pilot implementation also tends to build up resistance to the change. In Norway for example, the policy adopted in order to promote the introduction of groups into industry, was to form single groups in a number of different factories. It was expected that the advantages of these pilot groups observed in operation, would lead to the rapid extension of group methods in these factories. In practice this did not happen as the pilot groups built up a reaction in the rest of the factory against the 'privileged' members of the groups, and later by extension against the method itself. This effect has been called 'encapsulation'.

There seems very little doubt that the most successful general approach to the introduction of Group Technology involves comprehensive analysis in planning the change and comprehensive implementation.

6.2.(c) *Who should undertake the design and implementation exercise*
The third important policy decision to be made concerns who should undertake the group design exercise and plan and control implementation. It is worth noting here that nearly every successful application of Group Technology has been led and strongly supported by a senior director of the company. The changes required are company wide and there can be little hope of real success if the exercise is led by a specialist manager such as a production manager, for example, at a lower level in the organisation.

The director will, however, need someone or a team of people to do the required detailed planning and co-ordination. Three alternative approaches are possible.

(1) A company expert or team of experts.
(2) Outside experts.
(3) A company team plus outside experts.

The advantages of using a company team are, firstly, that its members are likely to have a better knowledge and understanding of the products made and of the processes used in the factory. Secondly, if they are involved in planning the organisation into groups, they are more likely to accept the change when it is implemented.

The advantages of using outside experts are that they are likely to have a wider experience of the problems involved in changing to groups than will be available inside most factories. The change to Group Technology is one which is likely to take place only once in the life of a company. There are therefore benefits to be obtained by employing an outside expert who has had experience of making this type of change in several different factories. The disadvantages of using outside experts are that generally they do not have a wide experience of the company's procedures, operations, products and problems.

The third alternative of using a company team plus outside experts is probably the best. It combines the experience of the outside expert with the local knowledge of the company's staff.

6.3. DEFINING OBJECTIVES

Before considering the problem of designing a group technology system, the objectives of the innovation should be defined and clearly stated.

6.3.(a) Advantages of a summary of objectives

The summary of objectives for the new Group Technology organisation, should list the main objectives which the system is required to reach. These may vary in detail from company to company, depending on the type of technological system, and on the type of market which it supplies. The summary of objectives will provide a basis for making decisions during the design and implementation stage. It is also useful during briefing and training, to have a clear statement showing why the changes are being introduced.

As an example, a summary of objectives was prepared for a company which machined parts in batches and assembled them to make a range of machine tools. Most of the products were of standard design but some variants of these designs were also supplied. These product variants were of three main types.

(1) Simple variants. These were made by adding additional parts to the standard products during assembly. Sets of parts for these variants were maintained in stock.

(2) Complex variants. These product variants required some special machining operations, as well as additional variant parts which were kept in stock in sets.

(3) Long lead variants. These were variants which required special high value purchased items. Delivery promises for these products included the forecast lead times to obtain these bought items.

6.3.(b) Example of a summary objectives

The proposed summary objectives for this company, with the units of measurement to be used for monitoring performance in brackets, was as follows.

(1) *Product design.*

    (a) Improved quality (number of inspection rejects).

   (b) Quicker introduction of modifications and new products (modification lead time).
(2) *Marketing.*
   (a) Reliable delivery against programme (number of overdue despatches).
   (b) Ability to accommodate variations in product mix.
   (c) Short lead time for changes in programme (weeks).
   (d) Reduced material throughput time giving shorter delivery times (weeks).
(3) *Production.*
   (a) Reliable completion of shop orders by due date (orders overdue).
   (b) Reliable quality (number of inspection rejects).
   (c) Reduced material throughput times (weeks).
   (d) Reduced stocks and work in progress (£).
   (e) Reduced setting up time (percentage running time).
   (f) Increased capacity (per cent).
   (g) Simpler production control (number of production control staff).
   (h) Increased material utilisation (utilisation per cent).
   (i) Reduced material obsolescence (obsolescence per cent).
   (j) Reduced number of operations per part (average operations/part).
(4) *Purchasing.*
   (a) Reliable forecast of purchase requirements (number of changes in annual production programme required per annum).
   (b) Minimum purchased item buffer stocks.
   (c) Reliable delivery from suppliers (number of overdue orders).
(5) *Personnel.*
   (a) Increased job satisfaction.
   (b) Improved motivation.
   (c) Reduced labour turnover (per cent).
   (d) Reduced absenteeism (days).

Most of the items in this summary can be quantified. Ideally they should be measured first under existing conditions, and then again at regular intervals during implementation and operation, to monitor achievement in meeting the objectives. Target figures can be adopted and, in fact, must be adopted if it is intended to do a feasibility study. It will be noted that these are total system objectives. Some of them, such as those under purchasing, are objectives which must be achieved before Group Technology can be efficiently applied, rather than expected advantages due to implementation.

It will be seen that some of these objectives are in conflict. For example item 3 under marketing, which calls for short lead times for changes in programme, must be limited if the objectives of production are to be achieved. In the case of this example, the final decision was to adopt period batch control with a period of four working weeks. Every four weeks there was a programme meeting at which the marketing department gave its sales forecast for the four week period starting twelve weeks ahead. For each of these programmes one four week period was allowed for data processing, one four week period was allowed for machining, and one four week period was allowed for assembly. Under these conditions the lead time for changes in the numbers of each product in the programme was 12 weeks. The lead times allowed for changes in the allocation of products to customers were shorter, however, as follows.

(1) *Standard products.* Lead time one week. The marketing department could change the customer allocation of any standard machine up to the day when it was completed and went for packing.
(2) *Simple variants.* Lead time two weeks.
(3) *Complex variants.* Lead time twelve weeks.
(4) *Long lead variants.* Lead time depended on purchase lead times.

6.3.(c) Feasibility study

Many companies will wish to undertake a feasibility study, to determine how much the change is likely to cost, and what financial benefits are likely to be achieved.

Methods of implementation are considered in the following chapter. There it is advocated that the total innovation should be broken down into a number of independent projects. Once this has been done it is comparatively simple to estimate the cost of introduction. Fairly reliable estimates can also be made of the periods

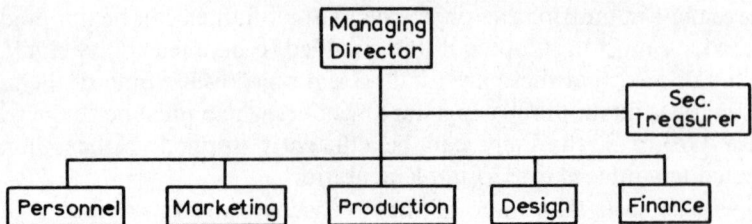

(a) *Divisional organisation*

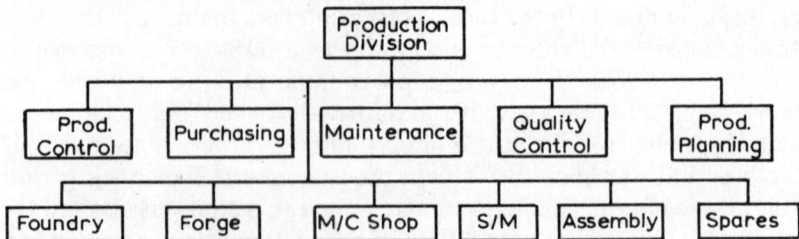

(b) *Organisation of the production division*

Fig. 6.1 *Organisation for Group Technology*

during which each project will be carried out and will incur expenditure. Providing that there is a quantified summary of objectives and a quantified appraisal of current operations, it is also comparatively simple to forecast the financial benefits which should be achieved.

It should be remembered, however, that such a feasibility study can only be carried out accurately after most of the preliminary planning has been completed. This means that much of the planning has to be based on an act of faith. It does mean, however, that there is a break point before the main expenditures are incurred, at which it is possible to pause and make sure that the effort will be worthwhile.

6.4. PRODUCTION SYSTEM SPECIFICATION

The next step after specifying the objectives, is to prepare a production system specification, giving details of the group technology system as it should be after implementation.

As an example, the following is a brief outline of the production system specification for the machine tool company described above:

6.4.(a) Divisional organisation
The final Group Technology divisional organisation is illustrated in Fig. 6.1 (a). The major changes introduced included:

(1) *Product design.* Changes in product design must be approved by the product design committee and can only be introduced by the approved modification and new product procedures, through the production division.

(2) *Marketing.* At the end of each four weeks period, marketing will provide a new short term sales programme for the period starting 12 weeks ahead.

At the same intervals they will revise the annual sales programme, adding one new period to complete the 13 period programme, and revising the remainder of the programme when necessary.

(3) *Financial control.* It is intended to discontinue the recording of operation times. The smallest cost centres in the factory, will, therefore, be groups.

(4) *Personnel.* It is planned to reduce the ratio of indirect and direct labour, particularly in the production division. The Establishment Committee will set the approved manning levels. The personnel division will ensure that retraining and natural wastage are used to reduce manning to these levels.

6.4.(b) Production division organisation
The organisation of the production division is illustrated in Fig. 6.1 (b).

(1) *Line departments.* The five line departments, foundry, forge, sheet metal, machine shop and assembly, will each be organised as an independent major group.

Each of these departments will be responsible for its own inspection, work scheduling, production planning, tool making, tool storage and maintenance and machine preventive maintenance inspections up to 'minor inspections'.

(2) *Production control.* The production control department will be

responsible for the preparation of the production programmes, the operation of production stores, the issue of shop orders to processing groups, interdepartmental materials handling, and the issue of stores requisitions for the issue of made and bought parts to assembly.

Ordering will be based on single cycle period batch control, with a four week cycle. Smoothing to give an even load on production will be based on the creation of a series of even-rate production programmes allowing some variations in finished product stocks.

Production control will be responsible for the maintenance of standard data, and for the accuracy of computer production files.

(3) *Purchasing.* The purchasing department will be responsible for maintaining a supply of all bought materials and parts, for the storage of such items and for their issue, against stores requisitions from production control, to the line departments.

They will base their purchasing on explosions and implosions from the annual production programme showing period requirements for one year ahead. These will be provided by production control at quarterly intervals, or more frequently, if there are changes in the annual programme. The purchasing department will be given target levels for maximum and minimum stock, for each major class of item.

(4) *Maintenance.* The maintenance department will be responsible for planning preventive maintenance inspections and will do major machine inspections. They will also undertake changes in layout, maintain stocks of machine spares, carry out major repairs, do any complete overhauls required, and maintain machine records.

(5) *Quality control.* The quality control department will plan all inspection procedures. They will also do sample checks on the work passed by groups in order to maintain inspection standards in the company.

(6) *Production planning.* The production planning department will be responsible for planning the division into groups and their modification as necessary. They will also control the allocation of parts to groups, the introduction of new processes, plant and equipment and plan the layout of plant in the factory.

(7) *Spares department.* The spares department will hold stocks of

parts for sale to customers as spare parts and will pack and despatch these items as orders are received.

6.4.(c) Organisation of groups
The smallest organisation units will be groups. These will be planned to exhibit the 'desirable characteristics of groups' (see Chapter 2). Each group will be independent with its own foreman or chargehand, reporting directly to the department manager.

(1) *Groups in line departments.* The processing groups in the line departments will be responsible for their own inspection, work scheduling, setting-up, and tool storage.

The production planning offices attached to each line department, will be divided into multi-disciplinary groups each responsible for methods, tooling and time standards in one or more processing groups.

Each line department will have a number of service groups covering common services such as material handling, tool making and lubrication.

(2) *Groups in production control.* Separate groups will be established for programming, standard data, ordering, stores and inter-departmental materials handling.

(3) *Groups in purchasing.* Separate groups will be established for the purchasing and storage of different classes of material. Initially these will cover raw materials, class A finished parts, other finished parts, and miscellaneous.

6.5. DEPARTMENTAL RE-ORGANISATION
One of the requirements for the successful formation of groups in a department, is that the department itself must have the characteristics of a very large group, or 'major group'. One cannot plan efficient groups in a department where many of the parts have to go to other departments or to outside sub-contractors for intermediate operations.

In some companies which carry out only one major process, this condition already exists and there is no need for further simplification. These factories are in effect already major groups. In other factories the organisation in departments may be basically satisfactory, in that each department completes the majority of components which it makes. The only problem here is to eliminate any exceptional parts which visit a number of different departments, or leave the factory for intermediate sub-contracted operations. This can normally be achieved by the redeployment of plant, method changes, redesign or by buying the part instead of making it. Finally there will be some companies where the division into departments has been based on minor process differences. In these cases considerable modification of the departmental structure will be necessary before it will be possible to form groups.

The Factory Flow Analysis (FFA) stage of Burbidge's Production Flow Analysis (16) provides a method of analysing interdepartmental material flow in order to simplify the material flow system. The aims of this technique are to establish departments so that:

(1) as far as possible each component is fully processed in one department only;

(2) as far as possible each machine type exists in one department only;

(3) departments have minimum valency, drawing their materials from the minimum number of sources and issuing them on completion of processing to the minimum number of destinations;

(4) incompatible processes are separated from each other.

These aims are incompatible and it is never possible to achieve them completely. They provide, however, a useful guide to the types of decision which will simplify the material flow system with the minimum investment in new plant and tooling.

Factory flow analysis is carried out in nine main steps as follows:

(1) *Find process route numbers (PRN).* Each existing department in the factory is given a number. The PRN is the code number formed by listing in correct sequence all the department numbers visited by a component.

(2) *Fill in the process route number frequency chart.* A chart is prepared showing the number of components which use each PRN code number.

(3) *Draw the system network.* A network diagram is drawn showing all the routes between departments along which components move.

---

(4) *Find the primary network.* The first step in simplifying the system network is to find the primary network. This network is formed using only those PRN's with the higher frequencies. Starting with the PRN with the highest frequency, new PRN's are added to the network until all departments have at least one arrow leading into them and at least one arrow leading out of them, or until a PRN is reached which imposes flow between two departments in both directions.

(5) *List restraints.* A list is made of any restraints which would prevent the joining together of departments. The main restraints will be concerned with incompatibility of process.

(6) *Simplify the primary network.* The primary network is simplified, first by joining departments together which are closely associated and for which there are no restraints preventing combination, and second by some redeployment of plant. This will produce a simplified network which will now accommodate many of the other PRN's not included when drawing the primary network.

(7) *Eliminate exceptions.* This simplification of the network will generally accommodate over 95 per cent of the parts. There will however be some exceptional parts which still cannot use the new simplified network. These exceptional parts must be examined individually. They can usually be eliminated by re-routing operations to other machines already existing in the major groups, a change of method, a change in design or by buying a component instead of making it.

(8) *Check machine loads.* The machine loads must be checked for any machines installed in more than one department so that the numbers required in each department can be determined.

(9) *Specify the standard inter-departmental flow system.* The specification of the final simplified inter-departmental flow system requires three main documents.

    (a) The simplified inter-departmental flow chart.
    (b) A list of plant in each department.
    (c) A list of parts processed in each department.

(a) *Original system network*

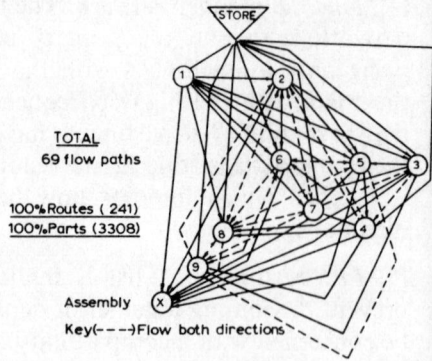

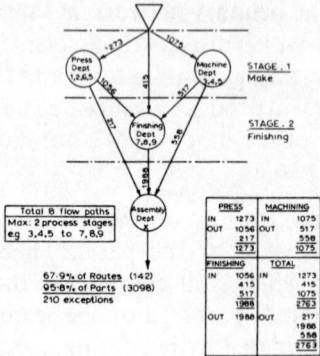

TOTAL
69 flow paths

100% Routes (241)
100% Parts (3308)

Key(- - -) Flow both directions

(b) *Primary network*

Total 8 flow paths
Max: 2 process stages
eg 3,4,5 to 7,8,9

67·9% of Routes (142)
95·8% of Parts (3098)
210 exceptions

	PRESS		MACHINING
IN	1273	IN	1075
OUT	1056	OUT	517
	217		558
	1273		1075

	FINISHING		TOTAL
IN	1056	IN	1273
	415		415
	517		1075
	1988		2763
OUT	1988	OUT	217
			558
			2763

(c) *Simplified network*

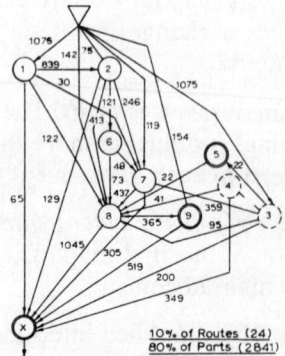

10% of Routes (24)
80% of Parts (2841)

Fig. 6.2 Factory Flow Analysis

An example of factory flow analysis is illustrated in Fig. 6.2. This shows the original system network, the primary network, and the final simplified network for a factory making mechanical calculators.

6.6. FINDING GROUPS

After establishing the departments as major groups, a choice can be made of the departments in which Group Technology is to be installed. Groups can be, and have been, installed in workshops with all types of technological system. This book is concerned, however, only with component processing departments such as machine shops and sheet metal shops. Generally speaking the division into groups in such departments is extremely complex, but the economic benefits which they give are greater than with other types of group.

The methods for forming groups in component processing departments can be listed as follows:

(1) by eye;
(2) classification and coding;
(3) production flow analysis.

Each of these methods will now be examined.

6.6.(a) Selection of groups by eye
The selection of groups by direct observation of the products to be produced and of the processes used to make them, is generally only possible in the simplest cases. It can be used for finding groups in assembly, in process industries, and in offices. In the case of component processing, however, it is only possible when the company makes a very small range of components. Although this method can be relatively successful when few parts are involved, the chance of missing components from their logical families is high.

The production engineer who does the analysis can use either design drawings, process planning sheets or actual components to help him in the analysis. If he uses component drawings, for example, he will attempt to stack these into separate piles which use different groups of machines, rather than relying completely on memory. Although this method has been used in one company manufacturing 500 different components, it is unlikely to be completely successful if more than 200 different components are manufactured.

6.6.(b) Classification and coding
The objective of classification and coding is to classify components by
their features and to code these features so that components having
similar code numbers possess similar features. There are three basic
features of a component which can be classified:

(1) shape;
(2) function;
(3) manufacturing operations and tooling.

Different classification systems use different features or combinations
of these features. There is a wide variety of systems available today.
In using classification and coding for group design, similarity in
features is used to group components into families.

The research covered by this book found 46 classification and coding systems. These are listed in Fig. 6.3. There are undoubtedly
many more. The difficulty with most of these systems, when using
them to form groups, is that they only combine some of the
components into easily recognised families and that they do not divide
the plant into groups of machines. This is so because most of these
classifications and coding systems are based on design features
found by examination of the component drawings.

Another difficulty is that they tend to bring together into the same
families, components which are similar in shape, but which due to
differences in requirement quantities or tolerances, should be made
on different types of machines and should therefore be made in
different groups. Conversely, classification and coding systems do not
bring together into the same families, parts which are dissimilar in
shape or function, but which should be together because they are
made on the same set of machines.

Name of coding system	Country of origin
1. OPITZ	W. Germany
2. BRISCH	UK
3. PERA	UK
4. VUOSO	Czechoslovakia
5. MITROFANOV	Russia
6. WILLIAMSON	UK
7. VUSTE	Czechoslovakia
8. KC—1	Japan
9. TOYODA	Japan
10. PGM	Sweden
11. NITTMASH	Russia
12. PITTLER	W. Germany
13. GILDEIMEISTER	W. Germany
14. STUTTGART	W. Germany
15. ZAFO	W. Germany
16. COPIC—BRISCH	W. Germany
17. IAMA	Yugoslavia
18. DDR STANDARD	E. Germany
19. HANIMAN GREEN	UK
20. VPTI	Russia
21. KOLAC	Czechoslovakia
22. STOCKMAN	W. Germany
23. CVM—TNO	Holland
24. WERNER and PFLEIDER	Germany
25. PERA SPECIALIST TOOL CODE	UK
26. LITMO	Russia
27. LANGE ROSSBERT	W. Germany
28. FOUNDRY CODE	Russia
29. IVANOV	Russia
30. BRUKHANOV and REBELSKI	W. Germany
31. KUKLEV	Russia
32. ANDREEVA	Russia
33. CZIKEL and ZEBISCH	W. Germany
34. PACYNA	W. Germany
35. GUREVICH	Russia
36. WALTER	E. Germany
37. AUERSWALD	E. Germany
38. PUSCH MAN	W. Germany
39. MALEK	Czechoslovakia
40. SALFORD	UK
41. ROMANOUSKII	Russia
42. KOBLOV	Russia
43. LABUTIN	Russia
44. VOSTRODOUSKII	Russia
45. GRIGOR 'EV	Russia
46. ODINTSOVA	Russia

Fig. 6.3 Classification and coding systems

Classification and coding systems based solely on design features tend to be efficient for variety reduction and design information retrieval, for which they were originally designed, but are not satisfactory as a basis for forming groups.

Attempts to expand the codes to cover production features so that they can be used to design groups have led to a large increase in the number of features coded. One example, the Miclass system developed by TNO in Holland, normally contains 30 digits. It can contain the information needed for production flow analysis, as well as information about design features. When used to find groups and families, it uses a combination of production flow analysis and design classification and coding.

There are also a number of classification and coding systems which have been designed specifically for non-machining processes, such as the Salford Mullary system for pressed parts and the Puschman and Opitz systems for sheet metal parts. In practice, there is considerable interaction between machining and non-machining processes and few classification systems can cope adequately with both.

One further disadvantage of classification and coding systems for finding the division into groups, is that they normally take a long time to install and their installation is costly. Because of these difficulties it is unlikely that any classification and coding system based solely on design features will be an effective method for finding a total division into groups and families.

6.6.(c) Production Flow Analysis methods of finding groups
Production flow analysis methods are methods of identifying component machine groups by analysing the machine to machine routes followed by all the components which are to be manufactured. All these methods are based on Burbidge's suggestion (16) that families of components and groups of machines exist naturally. It is the objective of these methods, by analysing the information contained in component route cards, to identify the best natural component machine groups, such that each group of machines will provide all the facilities needed to completely manufacture a particular family of components.

There are a number of different variants of production flow analysis now in use. In general they all involve the following main steps:

(1) refine the data;
(2) consolidate into sub-groups;
(3) combine sub-groups to form groups and families;
(4) load checking.

Steps 1, 2, and 4 normally require the use of a computer, but there is one variant for which there is a manual method. No one has yet succeeded in the complete automation of step 3. The final combination into groups and families is normally at least partly by eye.

6.7. REFINE THE DATA

The data required for production flow analysis, consists of:

(1) an accurate plant list;
(2) a route card for each component;
(3) a used-on record, generally in the form of a parts list for each product.

Most engineering companies can provide this information, but experience has shown that the level of accuracy of the data is generally low, and it is normally desirable to improve its accuracy before attempting to find the groups and families.

6.7.(a) The plant list

Most engineering companies have a plant list published for internal use, which shows all the production machines installed and in use. Generally these lists are only brought up to date at long intervals and it is rare to find one which is completely accurate. It will often be found that new machines have been installed which are not on the plant list, and that others are still on the list, which have been declared obsolete and taken out of production. The only safe way to correct these errors is to make a physical check of the plant.

Production flow analysis is simplified if the machines are classified by type in such a way that any part machined on a particular type, could be transferred to any other machine of the same type.

Normally this will bring together machines of the same make and model, but there are exceptions. For example a range of pillar drills of different makes but the same capacity, would be in the same class, and there might be two cylindrical grinding machines of the same make and model, which would be in different classes, if one was fitted with a wide grinding wheel and one with a narrow wheel.

6.7.(b) The route cards

Errors in the route cards arise mainly due to method changes, and to the transfer of operations from heavily loaded to lightly loaded machines. For load checking it is necessary to know the standard times for each operation. It will often be found that this information is missing from many of the route cards. It is desirable that most of these errors and ommissions should be corrected before starting production flow analysis.

If the routing data is on file for the computer, it can be programmed very simply to check the accuracy of the routes. Simple programmes can be prepared, for example, to find routes showing machines which are no longer on the plant list, to find routes without standard times and to check that there is a route for every made part in the parts list.

6.7.(c) Parts list

The parts lists are required mainly for load checking. Errors in the parts lists are a less frequent cause of difficulty than those in the plant list and route cards. What is needed is a reliable and systematic modification system to ensure that changes in design are immediately reflected by changes in the parts lists.

6.8. FINDING GROUPS – KEY MACHINE METHODS

The first step in most variants of production flow analysis is to reduce a large number of routes, to a smaller number of sets of routes, for which there is a high probability that all parts in each set must be in the same family. When the data has been simplified in this way, the final combination of sets to form groups, is always at least partly by eye. There are two main ways in which parts can be combined into sets.

(1) Key machine methods.
(2) Combination methods.

In this section two examples of the first method are described. Combination methods will be described in the next section.

6.8.(a) Group analysis

Group analysis is the second stage in Burbidge's production flow analysis. Based originally on matrix sorting by eye, the method was then developed for the computer (34). Once the method had been simplified for the computer, it was found that it could be used manually without difficulty (35).

The manual method is illustrated diagrammatically in Fig. 6.4. It follows eight main steps:

(1) list all the parts made on each machine type and count their number (F);
(2) find the 'nucleus' (key) machine, used to make the smallest number of parts;
(3) form a 'module' (set) containing all parts which use this machine and all other machines used to make them;

MACHINE USAGE SHEET									MIC TYPE n. C 16
Value of "F": 69									Sheet n. 1
MIC n.(s) : (3):C16·5: C16·6 and C16·10					Key: Part n. / Module n.				
Description : Gear hobbing m/c									of 1

1 10736	10737	11443	11462	11468	11500	11509	11722	11831	11869
1	1	2		3	4	3	3		2
2 12079	12080	12100	12119	12123	12186	12199	12221	12222	12365
5	5	4	4	4			5	5	
3 12372	12394	12395	12460	12503	12976	12978	13210	13276	13807
3	1	1	4	2	5	5		3	
4 13912	14011	14096	14097	14173	14174	14230	14231	14295	14311
	1	1	5	5	5	5	4	4	
5 14386	14404	14485	14492	14567	14902	14903	15627	15688	15721
		4		3	5	5		3	3
6 15729	15749	15789	15825	15964	15965	15999	16112	16665	16666
3	4	3							
7 16734	16798	16802	16811	16812	16864	16887	16895	16901	
	4	5	5	3		3			
8									
9									
10									

(a) *Machine usage sheet-all parts using each machine*

Fig. 6.4 Manual group analysis

(*Reproduced from* Production Engineer, *October 1977*)

			NUCLEUS SELECTION SHEET													Sheet n. 1 of 10	
Machines			MODULE N.														
			1			2			3			4			5		
Type	F	N.	f	εf	F εf	f	εf	F εf	f	εf	F εf	f	εf	F εf	f	εf	F εf
B	10	30	1			30			30			30			30			30
B	73	32	1			32			32			32			32			32
C	12	14	1	2	2	12			12			12			14	12	12	–
C	16	69	3	6	6	68	3	9	60	12	21	48	10	31	38	14	45	24
C	24	42	2			42			42			42			42			42
C	46	6	1	6	6	–												
C	58	12	1	12		12			12			12			12	12	12	–
C	68	36	1			36			36			36			36			36
D	11	30	1			30			30			30			30			30
D	15	227	12	5	6	221	7	13	214	10	23	204	14	37	190	10	47	180
D	34	302	10			302	1	1	301	3	4	298	2	6	296			296
D	42	56	2			55			55			56			56			56
D	76	15	1			15	2	2	13	13	15							
G	14	60	2			60			60			60			60			60
G	25	40	2			40	2	2	38	12	14	26			26			26
G	36	28	1			28			28			28			28			28
G	38	50	2			50			50			50			50			50
G	52	17	1			17	2	2	15	1	3	14	14	17	–			
L	12	180	6			180	8	8	172	13	21	159			159			159
L	14	205	6	5	5	200			200			200			200	14	29	186
L	15	46	1	5	5	41			41			41			41	12	17	29
L	·21	70	2	1	1	69			69			69	14	15	54			54
L	37	125	4			125			125			125			125			125
L	51	28	1			28			28			28			28			28
M	31	128	4			128	1	1	127	2	3	125	4	7	121			121
M	32	160	5			160			160			160			160			160
M	41	8	1			8	8	8	–									
M	62	54	2	4	4	50			50			50			50			50

Extension sheet n. — of —

(b) Nucleus selection sheet

Fig. 6.4 Manual group analysis
(*Reproduced from* Production Engineer, *October 1977*)

(4) split the module if the nucleus machine is used with two or more different groups of other machines and remove for individual examination exceptional parts which require the use of machines not otherwise used in the module;

(5) repeat 2, 3, and 4 until all parts and machines are in modules;

(6) combine modules to form groups;

(7) eliminate exceptional components by re-routing, changes in method, changes in design or by buying instead of making;

	MODULE SPECIFICATION SHEET									Module no. 6
	NUCLEUS M/C TYPE N. C16			DESCRIPTION : Hobbing m/c						Sheet no. 1 of 1

	PART Nos	C 16	D 15	D 34	G 25	L 12	M 31	L 14	L 15	L 21	
		69	227	302	91	180	128	208	46	70	
a	11462	✓	✓	✓	✓	✓	✓				
	13912	✓	✓	✓	✓	✓	✓				
	14386	✓	✓		✓	✓	✓				
	16665	✓	✓	✓	✓	✓					
	16734	✓		✓		✓					
b	16901	✓					✓	✓	✓	✓	
	12199	✓	✓					✓	✓		
	11831	✓	✓					✓	✓		
	12186	✓	✓					✓	✓		
	13807	✓	✓					✓	✓		
	14011	✓	✓					✓	✓		
	15627	✓	✓					✓	✓		
	15964	✓	✓					✓	✓		
	15965	✓	✓					✓	✓		
	15999	✓	✓					✓	✓		
	16112	✓	✓					✓	✓		
	16666	✓	✓					✓	✓		
	16798	✓						✓	✓		
	16889	✓						✓	✓		
	12365	✓	✓					✓			
	15210	✓	✓					✓			
	14404	✓						✓			
	14092	✓						✓			
	15729	✓						✓			
	TOTALS (f)	24	27	4	4	6	4	18	12	1	

(c) *Module specification sheet—all parts using nucleus machine*

Fig. 6.4 Manual group analysis

(*Reproduced from* Production Engineer, *October 1977*)

(8) check loads in each group imposed by a number of test programmes on all machine types required in more than one group. Plan division of machines of these types between groups.

The early modules produced by this method are formed round special machines and round general purpose machines which are used to make few parts. Typical examples are shaft centring machines, key seaters and gear tooth rounding machines. These machines are essential for the manufacturing of some parts, but are often very lightly loaded. Normally there will only be one of each of these early nucleus machine types. All parts using them must therefore be in the same groups to avoid the need to buy additional machines.

The middle range of modules is formed round machines which have already been used in some previous modules, and for which there are only a few remaining parts which use them. It is in this range that splitting of the module is most likely to be required.

MODULAR SYNTHESIS SHEET														Sheet n. 1 of 4
MACHINES			**MODULE n.:**												
TYPE	F	n.	2	3	4	6a	12	1	5	6b	16	18	15
C	16	69	3	3	12	10	6		6	14	18			
D	15	227	12	7	10	14	4	12	6	10	13	16	36	17
D	34	301	10	1	3	2	4	16						9
D	76	15	1	2	13									
G	35	40	2	2	12		4	22						
G	52	17	1	2	1	14								
L	22	180	6	8	13		6	6						15
M	31	128	4	1	2	4	4	10						20
M	41	8	1	8										
L	21	70	2			14		22	1		1			32
C	12	14	1						2	12				
C	46	6	1						6					
L	14	205	6						5	14	18	36	42	
L	15	46	1						5	12	12	17		
M	62	54	2						4			17	33	
C	58	12	1							12				
C	68	36	1									36		
C	24	42	2										42	

Extension sheet n. —— of ——

(*d*) *Modular synthesis sheet—combine modules to form groups*

Fig. 6.4 Manual group analysis

(*Reproduced from* Production Engineer, *October 1977*)

    Finally, as each module eliminates one or more machine types, the final modules are composed of simple parts made with very few operations on widely used types of machine such as lathes, milling machines and drills.

    When combining the modules to form groups, one objective is to have as many machine types as possible each existing in only one group. A figure of 80 per cent of machine types in one group only is

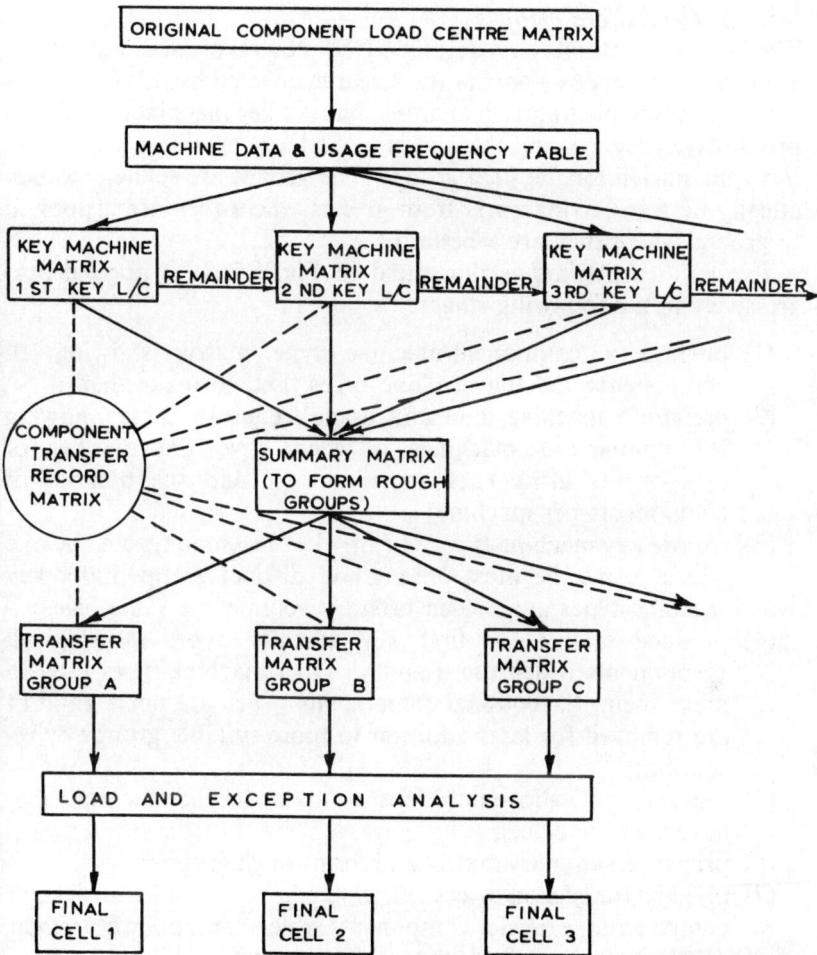

Fig. 6.5 The Salford method

typical, and it is mainly the common machines, of which there are
many of each type, such as lathes, milling machines and drills, which
are required in more than one group. This approach reduces the risk
that additional machines must be bought, simplifies load balancing,
and simplifies the manning of the groups.

6.8.(b) The Salford method

The Salford method was developed by Nagarkar and Norton at Salford University as a part of the research covered by this book. It is similar in principle to group analysis, but the key machines are found progressively by eye and the aim is to find key machines which will form the nucleii for the final groups. The groups are refined systematically by transferring parts from groups where they are a poor fit to groups where they are a better fit.

The Salford method is illustrated in Fig. 6.5. It is applied progressively in the following stages:

(1) prepare a component/machine type matrix showing all components and the machine types used to make them;

(2) prepare a machine data and usage frequency table, showing the number of machines of each type, the number of components using each machine type and the number of components per machine;

(3) choose key machine type. The first key machine type is the one which forms the most unique and distinct group. Later key machine types are chosen progressively on the same basis;

(4) produce matrix for first key machine type, showing all components which use it and all other machine types used to make them. Exceptional components which are not a good fit are removed for later addition to more suitable groups or rerouting;

(5) repeat 4 and 5 above until most components and machine types have been allocated;

(6) prepare a summary matrix and form rough groups;

(7) prepare transfer matrices on which to record all transfers of components. Transfer components which are a poor fit in their rough group to a more appropriate group;

(8) eliminate exceptions which do not fit into any group by rerouting, change of method, change in design, or buying the component instead of making it;

(9) analyse load for various demand levels to find the numbers of each machine type to be allotted to each group.

This method has been used successfully in the design of groups for a sheet metal working factory.

6.9. FINDING GROUPS – COMBINATION METHODS

Combination methods of production flow analysis start by finding all the different combinations of machines used to make components. These combinations are then combined to form groups.

The ratio of the number of parts to number of combinations is seldom more than 20:1, so these methods achieve less simplification of the data at the first stage of analysis, than the key machine methods.

Two variants of this approach will be described.

6.9.(a) Component flow analysis

This method was designed by Dr El-Essawy at Manchester University (UMIST). It has been used successfully to find groups and families in a number of different companies. The method of analysis can be summarized as follows:

(1) find all combinations of machine types used to make components and list the components using each combination;
(2) combine combinations using similar machines, by eye to form rough groups;
(3) analyse the sequence of machine type usage for all components in each group, and determine the load imposed by each of these sequences on each machine;
(4) determine the number of machines and equipment items to be installed in each group;
(5) plan the manning of the group;
(6) plan the layout of the group.

It will be noted that this method considers both combinations and usage sequences, and seeks to find not only the division into groups and families, but also the best arrangement for plant layout.

6.9.(b) Combination method

This method developed by Purcheck of the Cranfield Institute of Technology (36) again starts by finding all combinations of machine types, and then uses linear programming to find the best way to combine them to form groups.

Two different types of combination are recognised in this method. 'Host combinations' are those which cannot be combined with any other combination. 'Guest combinations' are those which can be combined with other combinations which contain all the same machine types. Most guest combinations can be added to two or more host combinations. A further degree of combination is possible by allowing a one machine difference and so on.

Linear programming is used to find the optimum way of combining combinations in terms of load balance and cost.

6.10. DEVELOPMENT OF THE GROUP

Before the groups can be installed, further planning is needed to determine the manning levels required, the layout of machines, tool storage arrangements and the methods of inspection and materials handling. Each of these will be briefly considered below.

6.10.(a) Group manning

The manning level for each group must now be determined. The load checks in machine hours carried out when planning the groups, give some indication of the number of man hours required. An examination of the types of machine allocated to each group and of the tolerances on the components in each family, will give some indication of the skill levels required. Allowances will have to be made where there are automatic machines, of which two or more can be operated by a single worker.

Labour flexibility can present problems in some companies. If a balanced work load is to be provided, it may be necessary for workers to move between different processes in the groups, requiring different skill levels. This will represent a change from current practice in most engineering companies today. Such proposals are not, in practice, totally unacceptable in the current industrial climate, but discussions with the shop floor personnel and with the unions at an early stage is vitally important. Job variety is becoming increasingly attractive to workers. A re-assessment of skill grades may be necessary under these circumstances but the additional labour cost will usually be balanced by the advantages obtained.

6.10.(b) Group management

The management of the groups must also be planned. It is highly desirable that each group should be independent. This means generally that they should have their own particular leader. If the groups are large enough, one foreman can be allocated to each group. With smaller groups a group leader or chargehand can be appointed to each group. In this case he will spend part of his time operating machines and part on administrative work.

6.10.(c) Plant layout

The first step is to plan the allocation of areas inside the work shop to each group, and to plan the relative positions of these groups. The positioning of the machines inside each group next involves consideration of such factors as the shape of the space available, material flow patterns, the positioning of storage areas, the type of part being processed and the volume of parts being processed.

There will inevitably be some constraints. Process plant, such as heat treatment plant, may be uneconomical to move. Heavy machines requiring foundations or pits may again be difficult and expensive to move. It is sometimes possible to plan the layout round these existing machines. The expense and inconvenience of moving such machines should be carefully balanced with the advantages in improved material flow which may result. In the UK, moving an average sized machine (say an 80 ton press or a capstan lathe), would currently involve a cost of around £100–£200. The actual cost of the move will depend partly on the methods used to hold machines to the floor of the shop and on the methods used to distribute electricity, compressed air and other services.

In determining the positions for machines inside the groups a main criterion should be the simplification of material flow. If the majority of components flow from one type of machine to another these should be laid out next to each other. If very diverse routes are found, the line analysis stage of Burbidge's production flow analysis (16) can be used to optimise the positioning of machines. This uses a flow chart method similar to that described earlier for factory flow analysis. In developing the layout, three dimensional scale models can be used. These enable all those affected by the change to study the new arrangement.

6.10.(d) Tool storage

It is generally desirable that all the special tools required by a group to make any component in their family of parts should be stored inside the group. Tools will then be available whenever they are required and considerable savings can be made in tool handling costs. If the group operatives have complete control of the tooling in this way, their attitude to the equipment used is likely to improve with a subsequent improvement in tool life.

6.10.(e) Quality control and inspection

Ensuring that an assembled product meets its specification is to a large extent a function of the quality of the component parts. The approach to quality assurance, quality control and inspection will vary, depending on the sophistication of the product being produced. In general, however, it is desirable that the responsibility for quality should be delegated to the groups. With simple products the operators can be trained to do their own inspection. With more complex products, trained inspectors may be required inside the groups, of which they should be members. Some companies have delegated inter-operation inspection to the groups, but have maintained a central inspection department for final inspection.

The responsibility for quality and inspection has been delegated to groups in many existing examples of Group Technology. The research has shown that in most cases this has led to an improvement in quality. As a part is normally processed completely in a group all the processes determining quality are under the control of the members of the group. They are in a much better position to maintain consistent quality than was ever possible with traditional process specialising sections. It is normally desirable to have a central quality control unit as well, which inspects samples of the finished work from all groups. Their main objectives are to maintain inspection standards in the groups and to advise on inspection methods.

6.10.(f) Materials handling

Materials handling procedures will have to be established for the groups. They include both procedures for supplying materials to the groups and for despatching finished work, and procedures for the handling of materials inside the groups. Each group should have its own storage area for incoming materials and outgoing finished components. Procedures should be established for handling materials to these storage areas and for clearing finished work from the group as soon as possible, after it is completed.

Inside the groups formal procedures for controlling materials handling should not be necessary. The group will move its own materials between machines as necessary. Simple handling equipment such as roller track, hand palete trucks or gantry cranes should be provided in each group to facilitate handling and reduce physical effort.

6.11. DEVELOPMENT OF RELATED SYSTEMS

To obtain the full advantages of Group Technology it may be necessary to modify many of the related systems existing in the factory. Changes may be required, for example, in product design procedures, sales and marketing procedures, production control, costing methods, personnel management policies, purchasing methods, and wage payments systems.

It is desirable that the necessary changes should be studied and planned before introducing groups and, in some cases, that they should be introduced before the change to group layout.

6.11.(a) Design procedures

The design department is unlikely to be greatly affected by the change to Group Technology. It is desirable however that they should understand the nature of the division into groups so that as far as possible they can design parts which can be completely manufactured in single groups.

It is desirable also that the methods to be used when introducing new products and modifications should be clearly laid down. With Group Technology there are frequent opportunities for the introduction of such changes. As far as possible, however, changes should be avoided during the life cycle of each short term programme. Frequent changes of these types only complicate the production system and make it less efficient. This is particularly important in the early days after introducing Group Technology, as the formation of groups will reveal many possibilities for variety reduction and will tend to generate a large number of modifications.

6.11.(b) Sales and marketing

The main requirement from sales and marketing is for regular and accurate sales programmes. Group Technology assists marketing by allowing the regular revision of their forecasts at short intervals, to accommodate changes in product demand. The provisioning of many of the bought parts, however, will have to be based on longer term, say, annual programmes. Short term variations, as reflected in the series of short term programmes, can be accommodated by carrying buffer stocks. In the long run, however, the efficiency of material supply will depend on the accuracy of the long term sales forecast.

Another way in which marketing and sales can greatly help production, is by discouraging specification changes after products have been ordered. Group Technology will normally give shorter and more reliable deliveries of finished products, and frequent opportunities for the introduction of modifications. It cannot accommodate the day to day changes in programme and product specifications which are common at present in many companies, any better than the existing traditional systems.

6.11.(c) Production control

The question of production control is discussed more fully in Chapter 9. Here it should be mentioned however, that this is one of the main changes required in most companies for the successful introduction of Group Technology.

The full advantages of group technology can only be obtained if all items required in each family are ordered together at regular period intervals. In many companies this involves a change from stock-control systems of ordering to flow-control ordering systems based on explosion from a series of short term programmes. It is often possible to make this change in the production control system before changing to group layout. This is probably the most desirable sequence of introduction.

6.11.(d) Costing systems

This again is a subject which is discussed in detail in Chapter 9. Several companies which have introduced Group Technology have treated their groups as individual cost centres. In other words they have eliminated operation time records inside the groups and treat the groups as though they were single machines. If this kind of change is envisaged, it is desirable that it should be completely planned before the change to group layout.

6.11.(e) Personnel management
Any necessary changes in personnel management procedures should also be planned before the change to group layout. For example, in some existing examples of Group Technology, the workers in each group have some say in the selection of new operators.

6.11.(f) Purchasing
In most companies purchasing methods will also need revision. The research has found that in many companies most deliveries from suppliers arrive late. In perhaps the majority of these companies the fault lies as much in their failure to give adequate notice of their requirements as with the supplier.

This subject is considered at greater length in Chapter 9. It should be noted, however, that no production system can work efficiently without materials. Some improvement in this field is usually desirable before changing to group layout.

6.12. CONCLUSIONS

This chapter has described methods of designing groups which have been found to be the most successful in practice.

Firstly, it has considered the policy for introduction, the specification of objectives, the new production system outline and the steps necessary to make an estimate of costs and savings.

Secondly, it has examined some of the changes which may be necessary in departmental organisation before groups can be designed. The technique of factory flow analysis which can be used for this purpose has been briefly described.

The different methods which can be used to find groups in component processing departments have been reviewed. The research has clearly indicated that the most satisfactory of these methods are those based on production flow analysis.

Consideration has been given to the development of the groups, including the planning of cell manning, cell management, plant layout, tool storage, quality control and inspection and materials handling.

Finally, the changes have been examined which may be necessary in design, sales and marketing, production control, costing systems, personnel management and purchasing to give the best results with group organisation.

CHAPTER 7

IMPLEMENTATION AND EARLY OPERATION

7.1. INTRODUCTION

This chapter studies the problems experienced with implementation and early operations by companies which have introduced Group Technology. It is hoped that a knowledge of the difficulties experienced by other companies will be helpful to new companies which make the change.

This chapter is based mainly on depth studies undertaken by teams from Bradford and Salford Universities in six companies which have introduced Group Technology, supported by research at Birmingham University and the London Business School. The chapter considers first the methods of implementation which can be used to introduce Group Technology. The research showed that the companies which have been most successful in introducing groups, have used the method of comprehensive analysis and comprehensive implementation, carried out by a company team with the help of outside experts, under the leadership of a senior director.

It is also submitted that it is possible to divide the total change to Group Technology into independent projects and to schedule implementation with the aid of Critical Path Analysis. With this information and a quantified summary of objectives, it is also possible to make reliable financial forecasts, for the control of the implementation.

The remainder of the chapter looks at the problems experienced during early operation of the new system. The research discovered that most of these problems could be traced to poor group design, inaccurate information systems, a lack of adequate training before the change and deficiencies in purchasing and production control methods.

7.2. IMPLEMENTATION POLICIES

When the division into groups and families has been planned together

159

with all the necessary changes in supporting systems, the next step is to implement these changes in the factory. By implementation is meant the actual physical changes in layout and the changes in procedures which are made when Group Technology is introduced. In considering implementation it is necessary first to re-consider the policy for introduction adopted at the design stage as this will, to a large extent, affect the approach to implementation itself. The actual methods of implementation will then be considered in the following section.

7.2.(a) General policy of implementation
In the previous chapter on the design of groups, a number of alternative approaches were presented. It was submitted that a comprehensive analysis, which considered all the parts to be made and all the machines which would be used to make them, followed by comprehensive implementation, had proved in practice to be the most successful approach.

It was submitted that the alternatives, of analysis based on samples and pilot introduction, could lead to difficulties such as the following:

(1) The need for special supporting systems for the groups in, for example, production control and payment methods, will make it necessary to have duplicated systems running at the same time.
(2) The selection of groups based on a sample of parts can lead to problems later when parts not in the sample are due for production. It also makes it impossible to check the load on the groups.
(3) When there are only a few pilot groups, the faster manufacture in the pilot groups may increase total inventories rather than reduce them.
(4) The formation of an apparently privileged pilot group may lead to a reaction against Group Technology which will make it difficult to introduce further groups ('encapsulation').

The findings of the present research support these doubts about the effectiveness of plans based on samples and about the use of pilot groups. With the introduction of new techniques for planning the design of groups and families, a total division can be planned relatively cheaply even when there is a large number of components. The apparent advantage of using samples therefore, that it reduces the complexity and cost of the design of the groups, is no longer valid.

The term 'comprehensive implementation' does not mean that the change has to be made instantaneously. Changes in layout and changes in procedures are obviously bound to take time. The term 'comprehensive implementation' merely indicates that once the company has decided on the total change, it intends to implement it completely over a period of time.

To simplify the management of change, there are arguments in favour of introducing one group initially to demonstrate how it works. This has been done successfully in one or two cases, but it is important that there should not be too long a delay between implementation of the first group and the implementation of following groups. In many factories there are already one or two sections which have many of the desirable characteristics of groups. Examples are a crankshaft section in a diesel engine factory, the tool rooms in many small and medium sized factories, and a biscuit cutter group in a factory which makes bakery machinery. If demonstration groups are needed, it is possible, with only minor alterations, to convert these sections into groups.

7.2.(b) Responsibility for implementation

With any major innovation it is desirable that an individual or group of individuals should be made responsible for ensuring that the change process is achieved satisfactorily. The present research has indicated, first of all, that implementation must be managed by some senior director of the company if it is to be successful. The changes required for a successful implementation of Group Technology affect all functions of management. It is essential, therefore, that someone with overall responsibility for management should be in charge of the change.

Reporting to this director, it is desirable that there should be an individual or group of individuals who are responsible for planning and controlling the implementation of the various projects involved in the change and for general co-ordination of the work. In practice this responsibility has been assigned in different companies to individuals, or small teams. Any outside experts appointed to advise on group design and methods of implementation will normally work with these special individuals, or teams.

7.2.(c) Involvement of company personnel

When introducing Group Technology it is beneficial to involve as many people in the work as possible, at all levels in the company, and from all functions. This will lead not only to a better understanding of the system but also to a greater degree of acceptance of the new system when it is introduced.

One method which has been used to involve large numbers of company personnel in implementation is to form temporary project planning groups. These are small inter-disciplinary planning groups containing members from all the functions affected by a particular change. Each group is given one specified project to consider, they have clearly defined terms of reference, and they are given a target date by which a solution is required. These project planning groups not only ensure that all aspects of a project are adequately considered, but they also have a valuable training function in that they familiarise large numbers of personnel with the principles of Group Technology.

One finding of the present research was that, in many companies which have applied Group Technology, the level of understanding of the objectives of Group Technology is generally low on the shop floor. This results from the lack of any training programme other than casual explanations. These findings in Britain contrast sharply with normal practice on the mainland of Europe. Companies which have introduced groups in countries such as Holland, Norway, Sweden and France have gone to great lengths to ensure that all company personnel are aware of the objectives of the proposed changes and how the new system will operate. It is strongly recommended that UK companies should follow this example.

7.2.(d) Time scale for the change in layout

It is not possible for a production system to be re-organised over night. In practice it will take months or even years to complete the change, but the time required to actually move the machines to their new locations can usually be measured in days. In a few small companies the change from functional layout to group layout was achieved during the annual works shut down over a period of two weeks. The most suitable period for modifying the plant layout is during this time when the works are closed for the annual holidays. Moving machines will inevitably cause disruption and the fewer people around at the time of moving large items of equipment, the more efficiently the task can be carried out. Where more time is required, further changes can be made during a succession of week ends. Each stage in this case should aim to complete the layout of one or more groups.

Most of the companies studied used their own maintenance department for the change in layout, supplementing it in some cases by a team from a company specialising in machine movements.

7.2.(e) Continuity of operations

In re-arranging the machines from a functional layout to a group layout, some loss in the continuity of operations must be expected. Losses in product output can be avoided by building up stocks of parts before the change, to compensate for the losses during implementation and early operation.

Immediately after implementation, a fall in output level can be expected while the company personnel learn to operate the new system. It has been found that the learning curve rises very steeply initially, output levels reaching the levels which were being achieved previously after four to eight weeks. The time required to reach the full potential efficiency of Group Technology can, however, be appreciable. This time can be reduced dramatically by recognising, during the group design and implementation stages, the difficulties which may arise, and removing the likely causes of inefficiency.

7.3. METHODS OF IMPLEMENTATION

Most of the companies which have introduced Group Technology have planned and implemented the change to group layout with commendable precision. The same cannot be said about the other necessary related changes. For the most part, the need for these changes has only been discovered after the change in layout. The complete change to a Group Technology system is extremely complicated. Experience has indicated however, that implementation need not proceed as a series of 'fire fighting' exercises. It can be planned with considerable precision.

7.3.(a) Implementation projects

When the Group Technology system has been designed, it is possible to divide the total list of changes required, into a number of independent projects. The aim should be to divide into separate projects in such a way that each project can be planned and implemented independently, and that each will bring some immediate benefit to the company, even if later projects have to be delayed for any reason.

Some of these projects will be planning projects. They will often include detail design of the system, as well as plans for implementation. Some of these planned changes will affect more than one function of management. These are particularly suitable for solution by project planning groups, as described earlier in the chapter. The remaining projects will be implementation projects. The aim here is to implement the plans made when tackling the planning projects. It will help if those who will have to manage the work after implementation are closely involved at this stage.

The advantages of this method are:

(1) each project can be implemented separately – there is no need to tackle all the problems of a major innovation at the same time;

(2) most of the planning projects and many of the implementation projects will have no effect on industrial relations;

(3) the few projects which may be difficult to introduce are easy to identify, are separated from other problems, and can be given special attention;

(4) each implementation project will contribute to the total change and also, in most cases, bring some immediate benefits to the company;

(5) as those who will have to manage the new system have been involved in its planning, they are more likely to accept the changes in practice.

As an example, Fig. 7.1 gives a list of the planned projects for the implementation of Group Technology in an engineering company.

1.2	Pareto analysis.
1.3	Correct routes and plant list.
1.6	Plan and introduce periodic programming.
1.8	Plan new wages system.
2.6	Revise buying methods.
3.4	Factory flow analysis (FFA).
4.5	Group analysis (GA).
4.6	Revise exceptional routes from FFA.
5.6	Plan 'period batch control' (PBC) method.
5.7	Revise exceptional routes from GA (*see* 4.5).
5.10	Plan manning of groups.
6.8	Plan changes in costing system.
6.10	Introduce PBC all made parts. 4 week cycle.
7.9	Line analysis (LA) for groups in Department W.
7.10	Tooling families for groups in Department W.
8.10	Introduce new costing and wages system.
8.12; 8.14; 8.16; 8.17	Change to group bonus in Departments W, X, Y, and Z.
9.10; 11.12; 13.14; 15.16	Plan layout in Departments W, X, Y, and Z.
9.12; 11.14; 13.16	Tooling families in Departments X, Y, and Z.
10.12; 12.14; 14.16; 16.17	Change layout in Departments W, X, Y, and Z.
17.18	Reduce PBC cycle to 2 weeks.

Fig. 7.1 List of implementation projects
(*Reproduced from Burbidge, J.L., The Introduction of Group Technology, 1975 (Heinemann, London)*)

7.3.(b) The schedule of implementation

The next step is to plan the logical sequence of implementation. For many of the projects the sequence will be obvious. For example, planning projects must be completed before implementation projects can start, and departmental reorganisation must precede the planning of groups and the change in layout.

If duration times are now estimated for each project, it will be possible to plan a schedule of introduction. As some of the projects can be implemented during the same periods of time, Critical Path Analysis is the correct technique for planning the schedule. An example of its use for this purpose is illustrated in Fig. 7.2. The duration times fixed for each project will depend partly on the number of

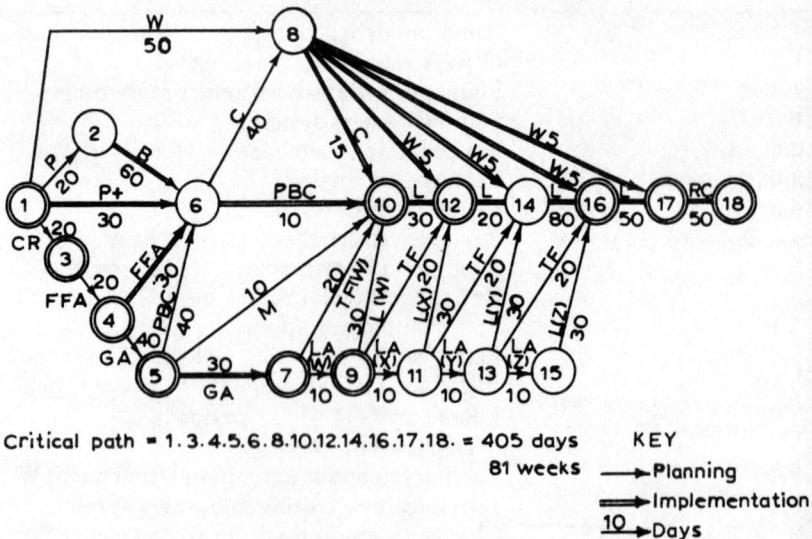

Critical path = 1.3.4.5.6.8.10.12.14.16.17.18. = 405 days

81 weeks

KEY

──▶ Planning

══▶ Implementation

10 ▶ Days

Fig. 7.2 Critical Path Analysis for scheduling implementation
(*Reproduced from Burbidge, J.L.*, The Introduction of Group Technology, *1975*
(*Heinemann, London*))

people and other resources assigned to them. This in turn will depend
on how quickly it is desired to complete the total change.

7.3.(c) Negotiations with unions

Some of the projects will require negotiations with the unions.
Examples are changes in method of payment, and changes in labour
flexibility. If the schedule of introduction is to be achieved, it is
important that these negotiations should start well in advance of the
proposed change dates.

To obtain the co-operation of the unions it is essential that they
should be kept in touch with developments. One way of doing this is
to hold regular briefing sessions, during the planning and implementation stages. It is also important, for good union relations, that
project planning groups should not be used for planning changes
which are normally the perogative of the unions and the subject of
collective bargaining. Unions are naturally sensitive on this point.

Group Technology has a good history as far as labour relations and
the unions are concerned. It has not proved difficult in the past to
convince the unions that the change is in the best interests of their
members. Many of those who have been involved in introducing

Group Technology have found that it was more difficult to convince their board of directors than the unions.

The most sensitive subject is likely to be the risk of redundancy for indirect labour. If there are prospects of increased sales, management can often eliminate this risk, by restricting new engagements early and allowing natural wastage, coupled with retraining, to effect the necessary reduction in numbers. Another way of reducing the risk of redundancy is to start new processing departments to make components previously purchased.

7.3.(d) Financial forecasts

One of the reasons why it has been difficult to persuade some boards of directors to introduce Group Technology, is that it is difficult to make reliable financial forecasts of the cost of introduction and of the benefits which will be achieved. This is, however, by no means impossible, once the initial planning has been completed.

The essentials for reliable financial forecasts are:

(1) a quantified summary of objectives (see Chapter 6), giving both present and target values;
(2) a list of implementation projects, with estimates of cost for each of them;
(3) a schedule of introduction showing starting and finishing times for each project.

The quantified summary of objectives provides a basis for the estimation of changes in both expenditure and investment. The main problem is to justify the target figures for the different objectives. Direct calculation can be used in some instances. For example, it is possible to calculate the effect of a proposed reduction in batch quantities on the stock investment. Inter firm comparison and comparison with the benefits obtained in other companies which have introduced Group Technology, can also be used.

The division into implementation projects and their cost estimates provides a firm basis for the estimation of the likely total cost of the change. Finally, the schedule of introduction provides a basis for estimating cash flows. The costs of the implementation projects can be allocated to different time periods without difficulty. The timing of the achievements of the economic benefits is more difficult to estimate, but again will be related to the times at which different projects are completed.

The cash flow forecast is unlikely to spring any unpleasant surprises. A study by the London Business School of seven companies which had introduced Group Technology found that in all cases the costs of introduction were recovered from benefits in less than one year. The result achieved by Serck Audco, where the costs of implementation were recovered four times by the reduction in the stock investment have been repeated in several companies.

Apart from providing figures which will help to convince boards of directors and accountants that Group Technology is worthwhile, these financial forecasts provide a means for controlling the introduction and will help to ensure that the benefits are actually achieved.

7.3.(e) Training
Training is an activity which experience has shown to be essential for the successful introduction of Group Technology. The first need is for briefing at all levels in the organisation, from directors to workers. The aim in this case is to ensure that everyone in the organisation has some understanding of the desirable characteristics of groups and of the objectives of the change. One company started with a weekend seminar for all managers, shop stewards and representatives of the workers from every department, and then followed with departmental meetings to examine the likely effects of the change on their work.

The next need is for the training of managers and supervisors, so that they fully understand how their duties will be affected by the new system. One of the best ways of doing this is to involve them in the detailed planning of the new methods. The use of inter-disciplinary project planning groups will give them a good understanding of the issues involved and of the way in which their individual jobs affect, and are affected by, the other functions of management.

For supervisors some technological training may be needed. A foreman for example, who has specialised in turning, may need some training in milling, grinding and drilling if he has to supervise these processes in his group. If the foreman is to be responsible for his own work scheduling, he should also be given some training in this work. Finally, in order to promote flexibility, facilities should be provided for the training of workers, so that they can learn to operate different types of machine.

7.4. PROBLEMS WITH EARLY OPERATION AT COMPANY LEVEL

After groups have been designed, the layout has been changed, and the system is operating, there is a tendency for management to regard the system as stable. The problems of operation are then left in the hands of first line supervision and shop floor personnel. It has been found, however, that for a long period after implementation, problems can arise at both company and group level which significantly reduce the efficiency of operation, but which are not recognised immediately as problems. It has been shown above that the time scale for the change in layout is normally short, usually of the order of days rather than weeks. In comparison, early operation, defined here as the period between implementation and steady operation, can often be measured in years. The problems experienced during this period will be considered firstly, in this section, at company level and secondly, in the next section, at group level.

7.4.(a) Information systems

The research has shown clearly the importance of accurate up to date information for successful Group Technology. This means two things. First the information system should be amended to recognise that groups exist, secondly the accuracy of the information files becomes more important and it should be improved as far as possible.

Some of the types of modification to the information files, which will be required are given below.

(1) *Planning file*. Route cards should be modified to show the number of the group in which each part is produced. The information which the route cards contain should be checked and corrected where necessary.

(2) *Plant file*. The group in which each machine or facility is located should be shown in the plant file. Lists of the machines and equipment provided for each group should also be made. The plant list should be physically checked to ensure that the machines on the list are actually installed in the factory.

(3) *Costing file*. The groups should be designated as cost centres and the costing system should be changed to relate costs to groups rather than to functional machining sections.

(4) *Control file*. The groups should be considered as control units, and control files should be amended accordingly.

(5) *Sales forecast file*. Information on future and current demand levels must be related to component requirements as well as product requirements. This is only possible if sales forecasts are produced systematically at regular intervals.

(6) *Design file*. The design file or used-on-record should show the groups in which components are manufactured.

Ideally the modification and updating of the information system should take place progressively during the design and implementation stages. In practice it has been found that although some information files may be updated, it is rare for the complete information system to be updated systematically. This has been identified by the research as a major reason in some companies for failure to achieve maximum efficiency with Group Technology.

Perhaps the most important information file from an operating view point is the planning file. In one company the route cards were not amended to show the group in which each part was made. This meant that it was impossible to calculate the work load on the groups and effective control by management was impossible. Only after operating the system for eighteen months, were the route cards amended to show the group in which each part was made. During this period the system operated well below its potential efficiency level.

7.4.(b) Production control system

The change from a functional production system to Group Technology generally requires some change in the production control system. The potential for improved and simplified control is considered in detail in Chapter 9. Here the discussion is limited to examples of what actually happened with production control, when the production system was modified.

Ideally, with Group Technology, the groups should receive a series of list orders, each covering one control period, at regular period intervals, and these should be accompanied by a load summary or load profile. This gives the cell foreman or leader the opportunity to plan the work to meet the order due-dates. It also makes it possible for him to plan the sequence of loading to minimise setting time. Finally, it gives him and the workers in the group, the opportunity to check that they are not being given an excessive load.

In most of the companies which have introduced Group Technology the importance of production control has been underestimated and changes have only been made after the introduction of group layout. This has meant operating for long periods below the level of efficiency which would otherwise have been possible.

In one company the production control system was modified, but only the first operation was considered in determining the work load which could be manufactured in a period. The capacity of subsequent operations was ignored. The problems imposed by this method might have been overcome if each component had only been routed to one group. In this company however the groups had not been well designed and 65 per cent of the parts visited more than one group. It was impossible to predict where overloads might occur.

In another company where the information files had not been updated, it was again impossible to plan a period load for the groups, because there were no means of identifying the groups on which different components should be loaded. This meant that planned work schedules gave way to control from shortage lists, the priorities on which were changing daily.

One of the primary objectives of an efficient production control ordering system in any company making assembled products should be to provide balanced product sets of parts for assembly. In many of the companies studied, this objective is left to the progress chasers. The rest of the production control system concentrates on 'loading' the machining groups. This inevitably leads to shortages, late deliveries, and a high investment in finished stocks.

Again many companies are still using multi-cycle ordering systems based on economic (sic) batch quantities. This usually means that the investment in finished stocks is high.

7.4.(c) Purchasing

If a given production programme is to be met, it is important that all the facilities needed to produce each batch of components, such as materials, tooling and manufacturing instructions, should be available when the job is issued to the work shop. The research has shown that material availability is a critical factor which, in many companies, limits the efficiency of Group Technology. Efficient procedures for ensuring that material is available when required should be developed and introduced before the change to groups.

In one very efficient installation of Group Technology, the purchasing procedures were completely revised before the change to group layout and some stocks of material were accumulated. This gave the groups a chance to start work under ideal conditions and prevented the disillusionment which is likely to arise if the groups can't work efficiently because they haven't got materials.

By contrast in another company it was found on analysis that 62 per cent of the material required to meet schedules was not available at the beginning of a period. Under these conditions it is impossible to run any production system efficiently or to control the level of work in progress.

In another company over 19 per cent of the shop orders were issued giving due dates which had already passed. Although it was recognised by all concerned that it was impossible to manufacture these orders by due date, they were a constant source of frustration. The problem in this and several other cases, was that the computer system for ordering was so designed that the issue of shop orders for processing and of material requisitions on the buying department for each component, were controlled together by the same data processing system. This meant that if they were to be completed by due date, shop orders using materials with long lead times would have had to be prepared a year or more in advance.

In one of the more effective production control systems, the procurement of materials is separated from the issuing of shop orders, and materials input from suppliers is regulated on the basis of 'explosions' and 'implosions' from the annual programme, showing period requirements. Buffer stocks are carried to accommodate variations between the annual programme and the series of short term programmes used to control processing in the work shops.

Most of the companies studied blamed their suppliers for bad deliveries. There was evidence, however, that many of them were themselves at least partly to blame. In some companies bad sales forecasts led to frequent changes in production programme. This coupled with inefficient production control systems meant that they did not themselves know what they needed until it was too late. It seemed unreasonable for them to blame their suppliers for not guessing.

7.4.(d) Work scheduling and progressing

Group Technology greatly simplifies the task of progressing work through the work shop. Because all operations on each part are carried out inside one group, it makes it possible for the group foreman to assume the responsibility for scheduling and progressing the work.

In many of the companies studied however, this advantage was not obtained. In one case, for example, progressing was the responsibility of the progress office, and the procedure for issuing work required an operator to report to the progress office for the issue of each new batch of work. This meant that the foreman had no control of the parts. The throughput histograms, shown in Fig. 7.3, demonstrate the effect of the consequent lack in continuity of operation. Histogram (a) shows how a part should progress through a group, but histogram (b) shows an example of the situation which often arose in the company, where after completing one or two operations on a component it was left as work in progress for a considerable time before the remaining operations were done. This gave rise to a situation where only 55 per cent of the shop orders were completed by their due dates.

Another recurring problem can be found in those companies which are attempting to use the computer for detailed work scheduling. In one company there are 70 people engaged on work scheduling alone, for a machine shop with 124 operators. To keep the system working the group foremen have of necessity been given authority to over ride the computer schedules 'where necessary'. One cannot avoid the opinion that in this case, and in most other cases of detailed computer scheduling, the group foremen on their own, could do the job more efficiently, at a tiny fraction of the cost. The main difficulty with computer scheduling is that it is difficult to maintain the accuracy of the information needed for efficient scheduling. Such essential data as the serviceability of machines and tooling, the availability of materials and operator absenteeism can change from hour to hour.

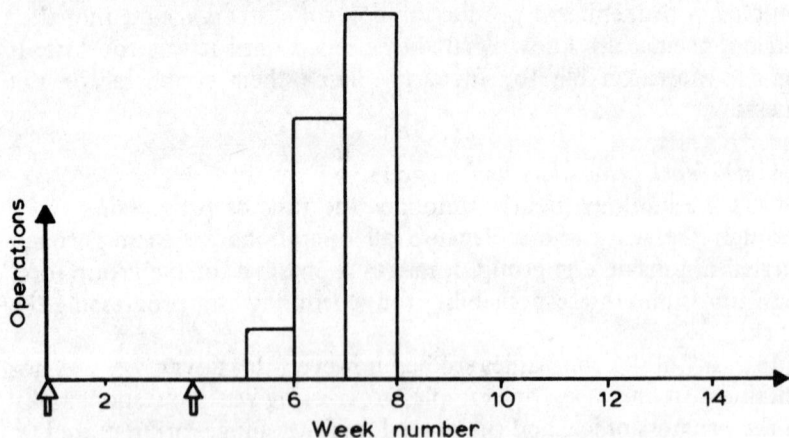

(a) *Potential throughput time with group control*

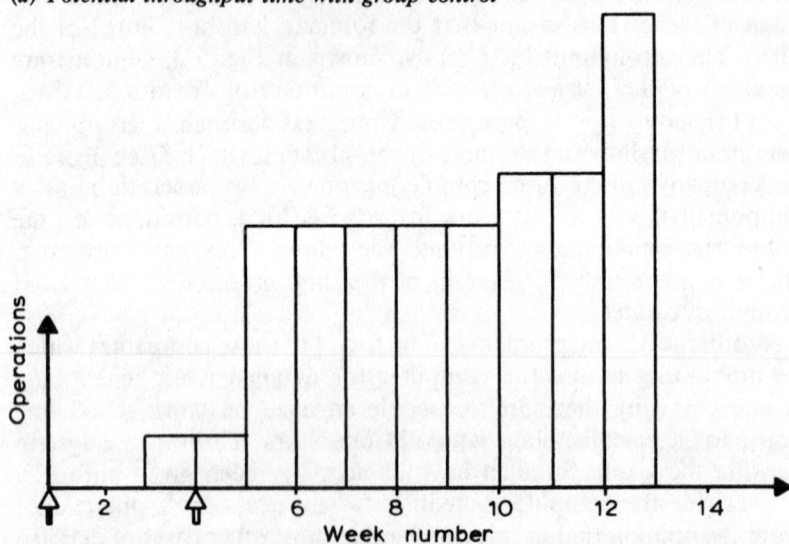

(b) *Actual throughput time with centralised work issue*
Fig. 7.3 Effect of centralised control on throughput time

The computer can help with the analysis of technical data, but scheduling is usually more efficient if the final responsibility is assigned to the foreman of the group.

In general the failure to change the production control system to suit Group Technology has been shown, in the majority of companies studied, to be a main reason for failure to achieve the full potential benefits of Group Technology. A strong case can be made that the change in production control should be planned and introduced either before or at the same time as the change in layout.

7.4.(e) Educational problems
There are several problems which arise at a personal and organisational level in the early operation phase of Group Technology. The majority of these problems can be traced to a lack of understanding of the objectives of Group Technology and can be classified as educational problems. These occur at both group and company levels. It was noted when discussing implementation that the involvement of company personnel in the process of change was often very low. This has also been found to be the case during early operation.

Even when the information needs of the production system from other company systems was clearly understood, it was found that these requirements were not being met. For example, it was found in several cases that sales departments were not providing sales forecasts in a suitable form or with sufficient accuracy and regularity. Under these conditions it was impossible to assess future material requirements and forward machine loads.

Even within the production function, a functional approach can still be found in the thinking of many managers after introducing Group Technology. The comment that 'Group Technology is a good approach, but it would be better if all the press brakes were together', is a type of comment which is repeated too often to be an exception. Machine dominated thinking in which the effectiveness of individual machines is given preference over the effectiveness of the production system as a whole, reduces the efficiency of operation.

7.5. PROBLEMS WITH EARLY OPERATION AT GROUP LEVEL
Having identified the types of problem which can occur at company level the difficulties which arise at group level will now be considered.

7.5.(a) Facility location
One of the desirable characteristics of groups given in Chapter 2 was

that the machines in each group should be laid out together in one area. This gives advantages in terms of reduced work movement which will minimise throughput times and work in progress and will also present the opportunity for the group to participate in the control of its own work.

The majority of companies which have introduced Group Technology have in fact created groups in defined areas. There is one example, however, of a company which did not change the layout of its machines when groups were introduced. Although a considerable reduction in setting time was achieved the reductions in work in progress and in throughput time were disappointing, and no social benefits were obtained.

In another company two groups were formed in different areas. An increase in the capacity of one of these groups due to the addition of a new machine, led to certain operations from the other group being transferred to the new machine. Although this maximised the utilisation of the new machine, overall production efficiency was not increased, due to control difficulties and delays in transferring work from one group to another. This is a case where it would have been better if the groups had been re-organised to cope with the increased capacity of the new machine.

7.5.(b) Work flow

Many people picture a group as a form of flow line. It is obvious that the simpler the material flow pattern inside the group the more efficient will be the operation of the group. In practice however in most groups different parts use different combinations of machines in different sequences. Relatively complex material flow systems generally exist inside groups, and a compromise layout must be achieved.

If the work flow is not carefully considered when planning the plant layout, long term problems may arise in operation. For example Fig. 7.4 (a) shows the work flow in one group in a particular company. Machine 1 has the smallest number of parts visiting it, but it is in a position where it interrupts the flow between machines 2 and 3. It would have been better to adopt the layout in Fig. 7.4 (b), which considerably simplifies material flow inside the group.

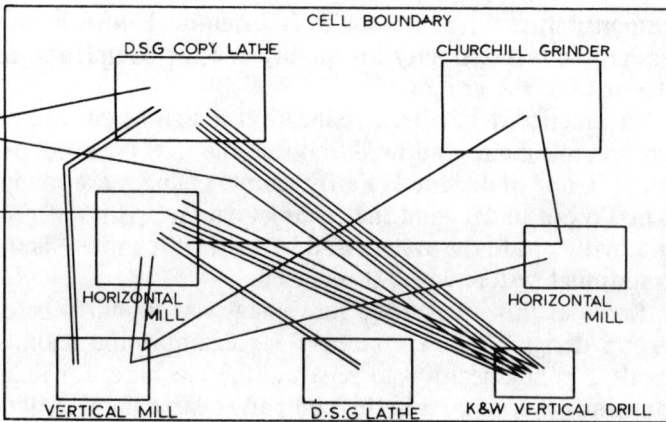

(a) *Existing layout*

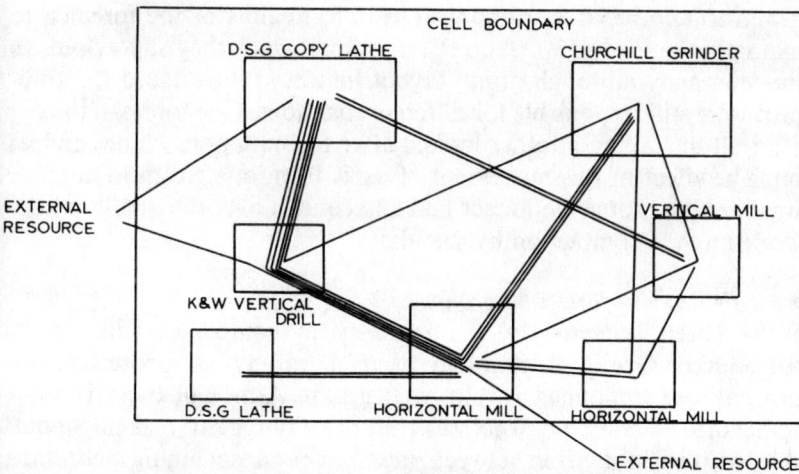

(b) *Improved layout*

Fig. 7.4 Layout of group for efficient material flow

7.5.(c) Group independence

Ideally each group in a Group Technology system should contain all the machines and equipment it needs to complete all the parts in its family. This is a necessary pre-requisite before the groups can achieve the degree of independence necessary for efficient delegation of responsibility. In particular it is essential before it is possible to assign the responsibility for quality and for completion of orders by due date to the groups.

In practice it has been found that a significant amount of inter-group movement is allowed to occur and that the effect of this movement is not understood. Fairly simple changes are made by people who do not understand the essential characteristics of groups, which gradually erode the system and make it little more efficient than the traditional system which it replaced.

Some of this inter-group movement was found to be due to poor group design. In one company for example the management had made a policy decision to separate the machines for machining and pressing processes. As in this company a large number of components require both processes most of the groups were not independent.

In another company many of the components moved from one group to another due to the well meant attempts of the foremen to maintain the load of work on the machines which they supervised. In this company although group layout had been introduced the foremen were still responsible for different functions. One foreman looked after all the lathes, another looked after the milling machines and so on. The effect of this movement of parts from one group to another was that the groups no longer had any control over the quality of the work or over completion by due date.

7.5.(d) Process variety in groups

It has been accepted by the majority of companies which have introduced Group Technology that a variety of processes will normally be contained within each group. Although there is some reluctance to combine processes traditionally regarded as incompatible, such as the division between press work and machining mentioned above, the advantages of establishing such multi-process groups will inevitably lead to the break down of this barrier.

Even where it appears that two machines are incompatible and should not be installed in the same group, it is often possible to take steps to prevent the detrimental effects of process interaction. For example, in one group making rotors and armatures for electric motors, the high speed press used to make the laminations was mounted on vibration absorbing mountings and acoustically

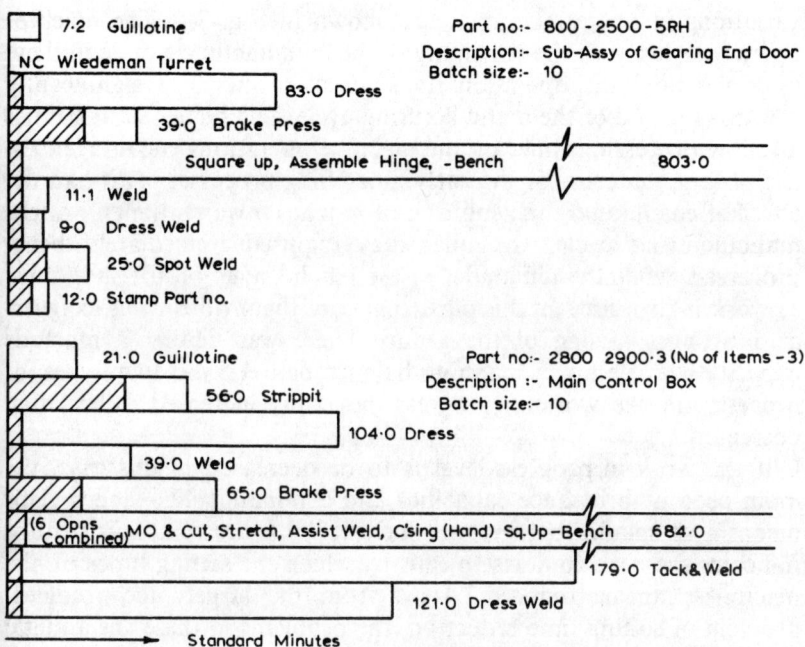

| 7·2 | Guillotine |
| NC Wiedeman Turret | |

Part no:- 800 - 2500 - 5RH
Description:- Sub-Assy of Gearing End Door
Batch size:- 10

83·0 Dress

39·0 Brake Press

Square up, Assemble Hinge, - Bench   803·0

11·1 Weld

9·0 Dress Weld

25·0 Spot Weld

12·0 Stamp Part no.

21·0 Guillotine

Part no:- 2800 2900·3 (No of Items – 3)
Description :- Main Control Box
Batch size:- 10

56·0 Strippit

104·0 Dress

39·0 Weld

65·0 Brake Press

(6 Opns Combined) MO & Cut, Stretch, Assist Weld, C'sing (Hand) Sq.Up –Bench   684·0

179·0 Tack & Weld

121·0 Dress Weld

Standard Minutes

Fig. 7.5 Examples of load balance in a group

screened so that it could be installed in the same group with the machine tools used for later operations.

7.5.(e) Group load balancing

In most machine shops, and also, therefore, in most groups, there will be some facilities which are heavily loaded and others which are only lightly loaded. In designing the groups an attempt is made to obtain the best balance of capacities between the groups. Any remaining imbalance between the loads on the machines in a group can only be balanced by allowing flexibility of labour so that the different processes can be operated as and when required. Failure to recognise this unavoidable imbalance will inevitably lead to problems.

In the sheet metal department of one company, for example, some of the groups carried out a combination of machine and bench operations. The ratio of machine to bench time was of the order of one to ten. Typical examples, showing the processes involved and the variations in operation times, are shown in Fig. 7.5. The machine operations occur at the beginning of the manufacturing cycle in these examples, and the operation time per piece increases significantly towards the end of the manufacturing cycle. The higher setting times and low processing times on the machine operations encouraged the use of large batches for the early machining processes. This had the effect of encouraging the splitting of batches towards the end of the manufacturing cycle, the quantities required immediately being processed, while the remainder of the batch stayed on the shop floor as work in progress. In this particular case the problem was exaggerated by poor design of the group. There was neither a machine capacity balance nor a manning balance between the two types of process and the work in progress inevitably increased rather than decreased.

If the work in progress level is to be decreased in this case, the main need is to balance capacities and eliminate split batches. This means that smaller batches will have to be processed. This in turn indicates the need to devise means by which the setting times on the machines can be reduced. Apart from the largely technological problem of setting time reduction, the problems in these sheet metal groups could have been avoided by better group design, improved planning of the work load, and increased flexibility of labour.

7.5.(f) Machine tools and equipment

When Group Technology is introduced it will be found that some of the machines in use are not ideal for their new role in producing a more limited range of components in small batches. It is important when Group Technology is introduced that machine procurement policy should be reviewed in the context of the new system.

Some of the problems which will have to be considered when framing such a policy are examined below.

(1) *More batches.* To obtain the full advantage of Group Technology parts must be machined in small batches at frequent intervals. Preference should be given therefore to machines which can be quickly set up.

(2) *Setting up.* To reduce setting times preference should be given to machines which can be adapted for quick change tooling, presetting and numerical control.

(3) *Common processes.* Some processes such as drilling, are likely to be required in nearly all the groups. For these types of process, simple and relatively cheap machines are better than very sophisticated, very high output and very expensive machine tools.

(4) *Number of operations per part.* Machines which can carry out a sequence of operations on components are more suitable for Group Technology than very high output machines which can only do one single process. It is probable that Group Technology will re-enforce a trend towards machining centres.

If new machine tools are obtained which increase the technological efficiency of a particular process, then the groups where they are used should be re-designed. To justify the purchase of capital intensive equipment, as much work as possible has to be loaded on to the new machine. This can severely affect the operation of the groups if they are not re-designed. One company which purchased technically sophisticated numerically controlled machines and installed them as substitutes for less sophisticated machines, found that in order to keep them loaded they had to transfer operations from a number of other groups. This eliminated most of the advantages of Group Technology. It would have been possible to re-design the groups to use these machines efficiently, but this was not done.

Another tendency noted in a number of companies which have turned to Group Technology is to keep obsolete machines as back up facilities in case of a break down. If this is done these machines should be carefully located so that they do not interfere with the efficient flow of materials inside the groups. In one company a large number of obsolete machines was left in the workshop and it was only after the groups had been in operation for 12 months that they were removed, providing valuable additional working space.

Obsolete machines can be valuable, in some instances, as a means of reducing setting time. They can be left permanently set up for operations requiring very long set ups, thus increasing capacity in the group.

7.5.(g) Machine utilisation

A reservation often expressed after the introduction of Group Technology is that it may appear to reduce machine utilisation. It should be remembered however that even in a functional system it is rare to find machine utilisation figures greater than 80 per cent and most machines will have a utilisation factor of 60 per cent or less. The change to groups will tend to increase the capacity of some machines by reducing setting times. It will appear therefore that these machines are being less fully utilised.

In the past the engineering industry has been obsessed with the idea that full machine utilisation is the secret of economic success. In following this aim other important factors have been neglected such as the investment in work in progress, efficient delivery to customers by promised delivery date and total product output.

One example of the strength of this traditional idea comes from the managing director in a company which had achieved a 20 per cent increase in sales output with the same plant and had increased its product output per employee by 50 per cent. He expressed the view that the main deficiency of Group Technology was that it had reduced his machine tool utilisation efficiency.

It seems certain that if the efficiency of production is to be improved it will be necessary to drop this one factor idea, that machine utilisation is the secret of economic production, from our list of over-riding beliefs. If the introduction of Group Technology can give significant increases in output per unit of investment and in output per employee, some apparent deterioration in machine utilisation can be allowed.

7.5.(h) Flexibility of labour

Some flexibility of labour is desirable in a group if it is to operate effectively. In practice it has been found that many companies are apprehensive about tackling this problem. To achieve flexibility the range of skills required by operators may need to be increased so that they can operate more than one type of machine. This may involve a re-assessment of payment grades to allow the changes to be made.

There are examples of companies where groups have been introduced without flexibility of labour and the system has nevertheless been able to operate efficiently. In the majority of cases, however, it will be found that if flexibility is not achieved, operating problems will result, particularly when variations in demand pattern are experienced.

The problem of introducing worker flexibility varies in different

		Summary of rejected and scrapped orders				
Month	Number of orders rejected	Reject orders as a percentage of total shop orders	Number of orders scrapped	Scrap orders as a percentage of total shop orders	Total of reject + scrap	Total as a percentage of shop total
January	56	7·46	15	2·14	72	9·6
February	66	8·80	36	4·80	102	13·6
March	69	9·20	31	4·13	100	13·33
April	67	8·93	36	4·80	103	13·73
May	62	8·26	23	3·06	85	11·33
					Average	12·30

Fig. 7.6 Low quality from groups without independence

parts of Britain. There is however a growing interest among workers and unions in the subject of job enlargement, and those companies which have made the effort have generally succeeded in negotiating an increase in flexibility.

In one company investigated during the research the opposite problem existed. In this company considerable effort had been devoted to de-skilling operations so that total flexibility of labour could be achieved. This had been done before the introduction of groups and management were anxious to retain the flexibility they had already achieved. When this company introduced groups, it did not assign particular operators to each group. The operators moved from group to group as new jobs were issued. Under these conditions control was to a large extent in the hands of the progress office rather than the foreman. It is inevitable in this case that quality will suffer and this in fact was found to be the case; Fig. 7.6 shows the percentage of shop orders rejected either for scrap or rectification.

7.5.(i) Mobility of labour
Although it is highly desirable that the teams of workers associated with each group should stay in their groups, there are circumstances where this may not be possible. For example, in a jobbing factory making different types of components in small batches by a number of different major processes, groups can be set up which are equipped to make different classes of item. This will give the normal advantages of Group Technology. It will reduce throughput time and enable the company to quote shorter deliveries. It will reduce the investment in work in progress and so on. In this type of company it is difficult,

however, to ensure an equal load of work, period by period, for all the groups. It may be necessary to arrange that some of the workers can be transferred from one group to another as the load varies.

In one of the companies studied, mobility of operators was needed in the welding department. One of the products of this department was manufactured only once or twice per year. When it was made it required a very heavy commitment from all the manufacturing resources. For a short period the majority of the workers had to concentrate on this one product. This meant a breakdown in group discipline which had to be accepted as long as these special products were produced. For the remainder of the year the advantages of group production could be realised.

7.6. CONCLUSIONS

The research has clearly indicated the types of difficulty which may be experienced in the implementation of Group Technology and in its early operation. It is obvious that constant vigilance will be needed if the full benefits of the new system are to be obtained.

The first point which comes out clearly is the importance of good group design and of arrangements for the modification of supporting systems where necessary. Without this solid basis of a well planned system, difficulties in implementation and in early operation are inevitable.

In a few cases, after months of planning and just before implementation, senior managers have made major changes, such as insisting that all NC lathes must be together and must not be in the same groups as other machine tools, which have largely eliminated the benefits of the change. In some cases this type of instant planning represented more a psychological need of the manager to show his authority, than any understanding of what the company was trying to achieve. Such changes should be resisted as far as humanly possible.

CHAPTER 8

PERFORMANCE MEASUREMENT

8.1. INTRODUCTION

The introduction of Group Technology brings with it a need for new methods of measuring performance. This chapter is based on a study undertaken by the London Business School. It starts by examining the methods of performance measurement which are currently used in industry and demonstrates why they are insufficient for controlling production and particularly for controlling the production from groups.

It is submitted that simple methods of physical performance measurement are essential for efficient Group Technology. Examples of the physical performance measures which are needed, include: the number of deliveries overdue to customers, the number of inspection rejects, the number of shop orders not completed by due date, and the number of overdue deliveries from suppliers. These types of information are seldom provided regularly and systematically in companies today. They are needed for the following reasons.

(1) They are needed by management, so that they can assess the performance of groups during operation.
(2) They are needed by management, so that they can ensure that the benefits of Group Technology are being achieved.
(3) They are needed by management, so that they can assess the effects of any changes they introduce in the production system and can control the development of the system.
(4) They are needed by the foremen and workers in groups, so that they can assess their own performance in meeting objectives and targets.

Existing methods of performance measurement are insufficient for these purposes, because they are concerned with a very small number of factors. Budgetary control, for example is mainly concerned with output and costs and gives little useful information about failures to deliver to customers on time, failure to complete shop orders by due dates, quality levels achieved and other factors vital to the well being of the company.

This chapter describes a framework of physical performance measures which could be used to meet this need. Methods of obtaining the data required and information about its use are also provided. Systems of performance measurement can vary from the very simple to the very complex. Examples are given at both these extremes.

It will be noted that the physical performance measures proposed in this chapter are very similar to the lists of quantified objectives examined in Chapter 6. Physical performance measurement can in fact be seen as a means for ensuring that the production objectives of a company are achieved.

8.2. EXISTING METHODS OF PERFORMANCE MEASUREMENT

Society in general today has an interest in the performance of industry. This interest is represented by shareholders and investors, management and other company employees, government at national and local level, ecologists, conservationists and many other groups.

It will be noted that there are two main levels at which company performance is of interest. There is one group of people who are interested in measuring the overall performance of a company by comparing its performance with that of other companies or with industrial averages. The second group consists of management and other company employees, for whom performance must be measured inside companies at a departmental, group, or individual level. This group of people will have some interest in comparing their performance with other companies, but will be much more interested in comparing their performance with past performance or with established norms, targets or budgets.

8.2.(a) Those interested in the overall performance of the company

The first group interested in performance measurement, consists of those who see the company as a complete unit and who are concerned with comparing its performance with that of other companies and with established standards. This group mainly comprises shareholders and investors, government at national and local level, boards of directors at local level and at head office, and the stock market.

For these people the prime interest is financial and for them the established performance measures are expressed in monetary terms. For their purposes performance measurement falls broadly into two groups:

(1) ratio compilation;
(2) factor productivities.

The first of these is the most used and well known. Comparison between the performance of different companies can be obtained by the methods of inter-firm comparison. A typical pyramid of ratios, as used by the British Centre for Inter-Firm Comparison, is given in Fig. 8.1.

8.2.(b) Performance measurement inside the factory

The second group of people requiring performance measurement data comprises managers and other employees within a factory, who are interested in the performance of the plant as a whole, of individual departments, of groups and of individual employees. For them the current methods of measuring and evaluating performance are less comprehensive than those for inter-firm comparison. In most factories today, the performance measures provided for this purpose can be divided into the three types; financial performance measures, individual man or machine output measures, and simple product output measures.

The most common method of financial performance assessment is the budget. This document is issued to each department and usually comprises a detailed break down of the costs which a department is expected to incur over the next period. Actual costs are measured and compared with the budgeted figures to obtain control. A departmental cost budget can provide an effective method for controlling costs and for providing standards against which performance in reducing costs and excess expenditure can be assessed. It will not show, however, whether the company is making the right products at the right time, whether they are maintaining the required levels of quality, whether they are completing the work they do by the planned due dates, or in general whether the department could have contributed more to the profitability of the company. The result of these features is that although budgetary control is useful in the appropriate circumstances, it is, on its own, basically punitive. It tends to show

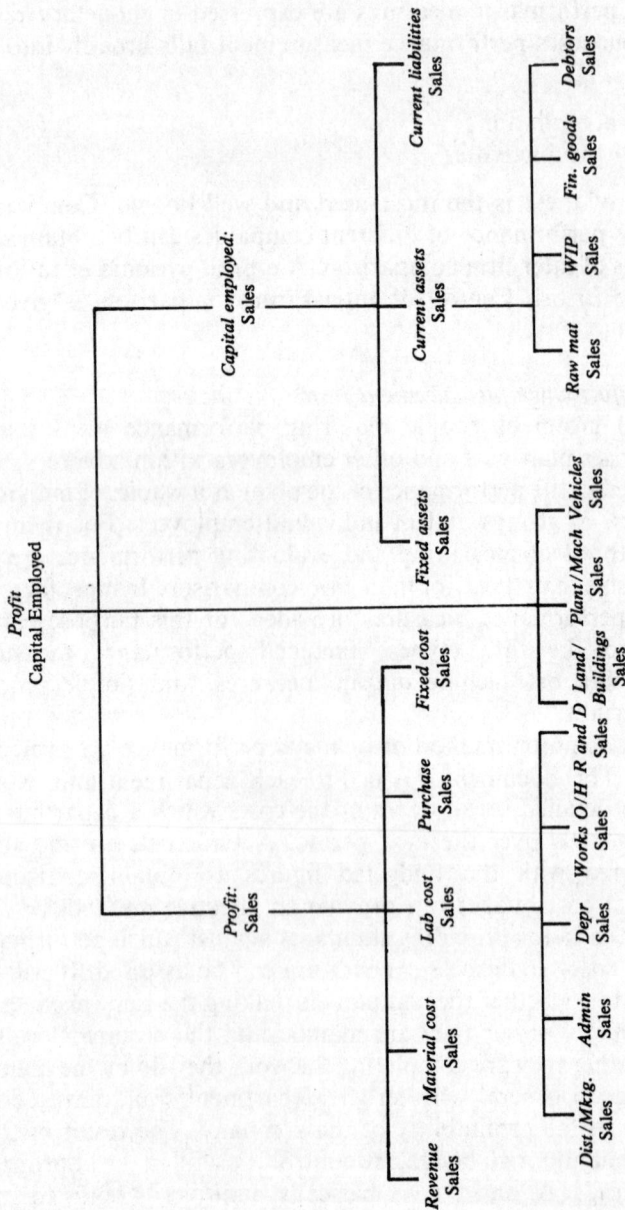

Fig. 8.1 Ratios used in inter-firm comparisons Performance Measures; top downwards system (interfirm comparison type)—an example of a 'conventional' analysis

deficiencies without indicating their causes, and without providing any useful help towards the design of a cure.

At a lower level in the organisation many individuals are judged by their output. Direct workers on the shop floor are set output targets above which they start to earn a bonus. Salesmen are often set sales targets above which they earn commission. Actual achievement is measured at regular intervals and is compared with those targets. These measures are designed to maximise the output of the company. They encourage output, however, as an absolute benefit, without considering whether the output is of the right quality or provides the product mix which will give maximum benefit to the company. Or whether, in fact, the components produced are wanted at all.

Output measures are applied to groups of workers in a similar way as to individuals. The targets for individuals and groups are set either by work study or from historical records of past output. Work study enables the effort of individuals, in producing a mix of different products or activities, to be translated into the common unit of standard hours.

Measures set by work study, in man or machine hours, are generally inappropriate for control at departmental level. Companies are therefore forced to use other measures at these levels. Typically the performance of departments and of whole plants are assessed by simple 'output' measures, in terms of value, weight, or number of units produced.

The use of a single measure to gauge performance encourages a distortion in factory operation because it is too simplistic. This may cause exaggeration of the importance of output at the expense of quality, profit, delivery and many other important factors.

8.2.(c) Deficiencies of existing methods of performance measurement
External performance is measured primarily for company share-holders, boards of directors and for external interests. It concentrates on financial issues. The various financial measures are well developed for inter-firm comparison, but as they are only of minor relevance to the present work, they need not be described here.

Internal performance measures are primarily used by managers and employees inside factories. They consist today, mainly of budgetary, work study and product output measures. Budgetary control is designed mainly to control expenditure. The comparison of operation times with work study standards is aimed mainly at reducing operation times, and the single physical output measures are too simplistic to be reliable.

The big gaps in between these methods of performance measurement have no adequate system of measurement. This is where the majority of managers operate and where most management decisions are taken.

8.3. FRAMEWORK FOR MEASURING OPERATIONAL PERFORMANCE

It has been demonstrated that over wide areas for which it has responsibility, management does not have sufficient information to control operations and to develop better policies and methods. This section of the chapter proposes a framework for measuring 'physical performance'. Each aspect of the problem is examined in turn. Methods and approaches are suggested, but the final details of the measurement system must be worked out on the basis of actual working experience inside each company.

8.3.(a) The scope of physical performance measurement

Physical performance measurement should provide a basis for the assessment of the efficiency of a company in carrying out the physical aspects of its production operations. The measures used should provide a basis for decision making in the following areas:

(1) as an indicator to show when things are going wrong and remedial action is required;

(2) as a means for monitoring the effects of changes in the production system;

(3) as a basis for the design of systems to increase motivation and reward;

(4) to assist in the integration of company activities;

(5) to provide feedback information to the foremen and workers in groups showing their achievement in meeting objectives.

A wide range of company activities exists about which management makes frequent decisions. These decisions affect such things as productivity, customer service, flexibility of production, quality, labour relations, and others. At present decisions are made in these areas without adequate information about existing performance levels.

Physical performance measurement is concerned with assessing the outcome of company operations in all these areas and with examining the actual achievement in operational terms at the different levels in the company. Control in these areas can be achieved by comparing the measures with forecasts, historical values and average industry values, in order to decide if each aspect of company operations is under control, or if it requires some management attention.

8.3.(b) The context of performance measurement

The context of physical performance measurement can be expressed as a set of fundamental assumptions about the operating management, and company environment.

(1) Performance is measured in relatively independent manufacturing units, e.g., a company, department or group which has the opportunity to perform, adapt, behave and take initiative on its own.

(2) Within the time span of measurement, the resources available to the independent unit remain fixed.

(3) Performance measurement should be related to a defined set of objectives.

(4) Planned levels of performance should be based on the constraints provided by the production system design.

(5) Performance is measured in relation to a set of activities, which involves a set of products, a market, sources of supply, a set of workers, a set of machines and equipment and a set of known standards.

(6) For any given production system design and current environment, there exists an optimal set of operating performance measures which will give the information needed to obtain optimal operational performance.

(7) Performance measures must be defined so that they create the opportunity to understand how to achieve the objectives.

(8) Production performance in the UK is low and the traditional accounting methods alone are not providing the means for raising performance because they do not relate to the pro-

duction system design nor do they indicate the means of making improvements.

(9) The selection of performance measures is critical. The stance we take on performance measurement reflects and summarises the values of our society.

(10) Performance measurement must relate to individual responsibilities, current patterns of control and to available data.

A number of important ideas arise directly from a consideration of these assumptions. Firstly, the independent unit of production will consist of a set of machines, labour and management which has a defined range of products to produce, a given target output and a given supplies condition. These are the conditions provided by Group Technology. Performance measurement is therefore likely to be easier and more productive with Group Technology.

Secondly, production operating performance can be measured in terms of the effectiveness in meeting targets and this is based on the effectiveness of the management of work flow in the use of the available capacity.

Thirdly, changes to the production system design, together with investment decisions, should be decided in the light of the external environment, should spring from the needs of the operating situation and should be reflected by recorded changes in operating performance.

Finally, implicit within all these assumptions and interpretations, is the fact that none of the measures are entirely independent of all the other measures. Each measure reflects some aspects of the physical operations of the company, which are themselves heavily inter-independent. For this reason all aspects of a company's operations should be monitored to ensure that no feature is allowed to dominate to an extent which detracts from the overall company performance.

8.3.(c) Performance measurement at different levels

Performance measurement is not restricted to any one level in a manufacturing company. This chapter, however, is concerned with performance measurement at one factory or site, so here this is the highest level for consideration. The department or workshop is the second level and the group is the lowest level.

These levels although similar in structure for measurement purposes, can be viewed as contributing at three different levels in the hierarchy of planning and control functions within a company.

(1) *Strategic planning.* Long range planning based on comparing the firm with its competitors, or related firms.

(2) *Tactical planning.* Broadest level of planning and control of the firm in terms of its broad functions, products, technology and organisation.

(3) *Operations management monitoring and control.* Planning and Control of operations in individual workshops and groups.

At each of these three levels it is possible to estimate the potential for any aspect of company operations, measure the actual performance, and in suitable cases recommend some remedial action. In each case it is important to be able to assess the value for any measure as good or bad, acceptable or unacceptable. The major difference between the levels is in the time scale and detail of the measurement taken. For strategic planning purposes, annual or quarterly analysis should be sufficient but for detailed operations management, weekly analysis may be necessary. The careful control of this aggregation of physical and time detail at each level, is fundamental to the success of performance measurement. Too much detail means too many measures and greater expense. Too little detail means providing little help to management.

8.3.(d) The control of performance factors
One of the difficulties faced by a production manager is to separate the variables which can be controlled and for which he is responsible, from those which are essentially fixed for a year or more and also from those which are controlled by the firms environment. There is a tendency to avoid this definition of the problem and so see the operations management task as one enormous complex problem.

The need is to separate the day-to-day or week-to-week operations management variables, which an immediate management decision may influence, from those which are essentially system design variables, which take longer to change. Similarly it is necessary to separate the system design variables from the environment variables, which are even slower to change and even less under the manager's control.

Material	Environmental variables	System design variables	Operating variables
OUT	Sales market Competition	•Product range •Product prices •Marketing policy	•Production programme 0 Product deliveries
IN	Supplies market Labour market Money market	•Purchasing system •Make or buy policy •Employment •Machines, tools, and equipment 0 Stocks and WIP	•Purchase orders •Delivery schedules •Sub-contracting •Overtime 0 Cash in bank
TRANSFORMATION	•Site •Buildings Technological constraints Political constraints Legal constraints	•Organisation •Layout •Production control methods •Production planning •Costing system •Payment system •Personnel policies 0 Capacity 0 Profit	•Manufacturing orders •Batch quantities •Labour allocation to jobs •Machine allocation to operations •Method changes •Design modifications 0 Cost 0 Output of products and parts 0 Material scrap
TIME OF CHANGE	YEARS	MONTHS	WEEKS OR DAYS

Key: (•) Parameters,   (0) output variables

Fig. 8.2 Examples of environmental system design and operating variables

8.4. THE PERFORMANCE MEASUREMENT VARIABLES

A suggested division between these three sets of variables is given in Fig. 8.2. This is not the only way that these variables could be grouped. The importance of any variable varies from company to company, and some of the allocations made might be made at different levels in different companies. It should be noted too, that some of the variables can appear in two or more of the divisions.

8.4.(a) Classification of variables
The variables in Fig. 8.2 have been divided into the following.

(1) *Environmental variables.* These variables take a long time to change and are largely outside the control of the company.

(2) *System design variables.* These variables can be controlled by the company but also take a long time to change.

(3) *Operating variables.* These variables take a short time to change and are more directly controllable by the production management.

The content of these different classes of variable will change at different levels in the organisation. At each level all variables with a time scale for change which is equal to or less than the time scale of the analysis, are controllable variables and those with a longer time scale may be considered as environmental variables. For example, at the lowest level in the organisation, the process routes are a fixed feature of the manufacturing system, but for medium and long term analysis at higher levels, they become increasingly variable.

At any given level, the controllable variables can be characterised by performance measures, but the environmental variables can only be measured. For them there is no issue of performance, good or bad.

8.4.(b) The choice of performance measurement factors
The variables shown in Fig. 8.2 are in most cases combinations of a large number of sub-variables. It is possible, however, to select a few significant sub-variables, or factors, which will be sufficient to provide a basis for control.

If the number of factors chosen is very large the cost of the analysis will be very high. The need is to find a compromise between cost and comprehensiveness, and at the same time to provide a simple scheme which may be built onto when more experience is gained.

A further criterion which should be followed in the choice of factors is that for the performance factors the measures must be within the control of the manager concerned. The operations manager, for example, does not control prices and so cannot be responsible for profit but only for costs.

8.4.(c) The list of physical performance factors
A list of physical factors considered important for operations management in one company, is listed in Fig. 8.3. The factors listed under 'system characteristics' are those which are either outside the control of the operations manager or those which would take a long time to change. Some of these need only be measured at fairly long intervals. For others, values would be required every period.

The 'operating variables' on the other hand are mainly perform-

System characteristics	Operating variables
•PRODUCT RANGE	•PRODUCTION PROGRAMME
No. of products	Products programmed
No. of variants	Products completed
Components/product	•PRODUCT DELIVERIES
No. new components pa	No. delivered to customer
No. modifications pa	Average delivery time quoted
PURCHASING	Total overdue orders
No. of suppliers	Average time overdue
No. bought items	Returns—defective
No. subcontracted ops	Spares delivered
•EMPLOYEES	Spares orders overdue
No. of employees(All)	•PURCHASING
No. direct workers	No. purchased orders issued
No. supervisors	No. purchase orders o/s
No. office staff	No. overdue
Others	Average time overdue
•MACHINES	Average lead time
No. of machines	Quality rejects
•INVESTMENT	•WORKSHOP ORDERS
Value. machines	No. parts ordered
Value. ancillary equipment	Average batch quantity
Value. raw materials	No. not done by due date
Value. work in progress	Quality rejects
Value. Finished products	•EMPLOYEE PERFORMANCE
Other assets	Labour turnover
Total value of assets	Labour absenteeism
•ORGANISATION	Accidents
No. of groups	Output per employee
Workers/group	Overtime hours
Machines/group	Period load (man hours)
Components/group	Load/capacity (per cent)
Workers not in groups	•MACHINE PERFORMANCE
•PRODUCTION PLANNING	Machine down time
No. of made items	Period load (m/c—hrs)
No. ops/component	Load/capacity (per cent)
No. method changes	

Fig. 8.3 Examples of physical performance factors

ance factors which can be measured, and are controllable, by the production manager. It will be noted that they will provide information about product deliveries, spare parts deliveries, purchasing, work shop orders and employee and machine performance, which are not at present normally provided systematically in most companies.

This list of physical performance factors might form the basis for a periodic report on the performance of the operations management. The essence of the system is that the system characteristics describe the environment in which the manufacturing manager works and incorporate previous manufacturing policy decisions and objectives. The performance factors indicate how the system is operating. By considering both these sides of the question an assessment can be made of the efficiency of operation and of the likely effects of change in the system.

8.4.(d) Setting performance standards

A fundamental problem in the application of any measurement system is the setting of standards of performance. In setting targets for these performance measures it is vital to recognise and understand the interaction between them. For example, to obtain maximum material flow rates it will help to have surplus capacity. Equally, to use capacity fully it will help to have plenty of work in progress and a poor delivery performance. It is impossible simultaneously to maximise all factors. The need is to choose the set of factor values which are potentially the best that can be chosen for a particular physical and policy situation.

Within this limitation that the different factors are closely related, there are two approaches to assessing a set of performance measures.

(1) Is production improving? In other words, are the current values better than last months?

(2) How far do the measures vary from some set of standard or potential values? The target or standard values can be obtained from a comparison of industrial achievements in the same sector of industry, considering average and extreme values, or from consideration of the technical and physical constraints alone. At the moment neither of these methods are

applicable for many of the measures, either because the data is not available, or because the constraints are not understood.

In practice the first method given above for setting target values is the most practical. The values for the performance factors recorded each period can be compared with such values as:

(1) the values for the previous period;
(2) the value for the same period one year previously;
(3) the average value for the last 12 months.

The alternative is to set arbitrary values based on policy requirements.

8.4.(e) Measurement of group performance

It is important to judge whether a group is achieving what was expected of it and to monitor its performance in some way to see that it continues to perform well. It is also important that the foreman and workers in the groups should get regular information about their performance and should be able to check their efficiency in achieving the group objectives.

This measurement of the performance of groups is made more difficult by the objective to reduce paper work in the group in order to increase productivity, and to eliminate the deleterious effect that direct and obvious management interference and 'over seeing' would have on the group. Fortunately a large part of this problem can be avoided by measuring the group performance from outside the group. Figure 8.4 shows a list of the physical factors which might form the basis for a regular period report on the performance of a machining group.

In this figure the values of the system variables, over which the group can have no control, have already been entered at the beginning of the period. The performance measures over which the group do have some control will be entered later at the end of the period. They include such factors as attendance, scrap and over due orders. All this information can be obtained from outside the group. It is necessary that the input of material should be carefully measured by the stores. The output of the group should also be carefully measured at a 'quantity control point' to which all deliveries from groups are routed. This same quantity control point can also check the scrap

Performance measurement — 199

Group performance	Group No. M6		
NUMBER OF PARTS IN FAMILY: 302	PERIOD: 22		
A. *Orders for period*	No:	%	Standard
1. No. different part nos. (% of family)	258	85	80%
2. Total No. parts	10 720		
3. No. different m/c operations	1 295		1300
4. Average ops/part	5		
5. Total load: m/c hrs.	1 208		
6. Total load: man hrs.	1 276		
B. *Men*			
1. No. of operators (direct)	20		
2. Man hours capacity	1 600		
3. Absent hours (all causes)			5%
4. Overtime			5%
5. Available man hours (% of 2 above)			100%
6. Load ratio (load man hours/cap) A6/B2	—	79	80%
C. *Machines*			
1. No. of m/cs	32		
2. Setting and ancilliary time			35%
3. M/c hrs. capacity (nett)			65%
4. Idle time hrs.			10%
5. Available m/c hrs. (% of 3 above)			90%
6. Load ratio (load m/c hrs/cap) A5/C3			75%
D. *Materials*			
1. Received (units=ea)	10 490		
2. Scrap			2%
3. Utilisation %			98%
4. Late delivery to group	230	2	3%
E. *Output for period*			
1. No. of parts overdue (end of period)			3%
2. No. parts nil delivery (end period)			0

Fig. 8.4 Measuring group performance

values, by noting the difference between the quantities of components received and the quantities of materials issued.

8.5. A PRODUCTION RECORD ANALYSIS PROGRAM (RAP)

The previous section of this chapter described a simple method for recording group performance from outside the groups. It involved the recording of the minimum amount of data and would be easy to install and use. Work has also been done at the London Business School to develop a more complex system of records together with a computer assisted procedure for analysing and assessing the data. This has been christened the Record Analysis Program or RAP.

In this work, data was required from the groups showing the times at which operators started and finished each operation and the machine on which it was done. This information was obtained from batch history cards, which are completed inside the groups by the operators as they do each operation. The batch history cards are collected at the end of each period for analysis.

RAP is a computer package which is available in both off-line and on-line forms. The package accepts data in a form which can be specified by the user and possesses the ability to carry out any or all of a range of analysis. There are seven analysis types built into the system ranging from simple counts, through cross tabulations to statistical computations, any of which may be applied to any portion of the data and to specified variables in the record. The computer produces a graphic output, the form of which can be selected for easy reading within the bounds set by the analysis. The RAP program has been tested in eight different companies. Four of these had traditional forms of organisation, and four were organised in groups.

8.6. CONCLUSIONS

Traditional forms of performance appraisement are based at one extreme on financial budgets and on comparisons of the actual time taken to complete operations, with the standard times found by work study. At the other extreme they are often based on simple product output measures. These methods do not provide management with all the information it needs to manage production.

In particular, present methods of measuring performance fail to provide management with any systematic information about such important facets of company efficiency as delivery performance, purchasing efficiency, the efficiency of workshops in meeting their commitments, employee performance and machine performance. Some of this information is available in some companies, but there is no systematic method of presenting the information at regular intervals to production managers, so that they can base their decisions on facts rather than guesses.

Particularly, with the introduction of Group Technology, methods are needed for assessing and controlling the performance of groups, and for providing feedback to the groups themselves, so that they can assess their own performance.

This chapter has described a method developed by the London Business School for building up a system of physical performance measures. Such systems can be either very complex or very simple. In either case, they fill an urgent need by supplying production management with the information it needs to make reasonable decisions.

CHAPTER 9

CONTROL

9.1. INTRODUCTION

One of the important findings of the research on which this book reports, is that one cannot obtain the full advantages of Group Technology in most companies without first changing several of the supporting control systems to suit the new type of organisation. The methods used at present for product design, production planning, production control, purchasing, cost control, wage payment and supervision, will have been designed to suit the original traditional form of organisation. Unfortunately such systems tend to be inefficient when used with Group Technology, mainly because they were not designed for that type of system.

This is particularly true of production control. The methods here are so important for successful Group Technology, that some authorities look on them as an integral part of Group Technology and not as one of the supporting changes.

This chapter is, therefore, mainly concerned with the methods of production control necessary to obtain the full advantages of Group Technology. It attempts to show why traditional multi-cycle systems do not work efficiently with Group Technology, and to describe the elements of the production control system which has been found, in practice, to be most efficient with groups.

The chapter also looks at the closely related subject of purchasing. A number of cases were discovered during the research where Group Technology was failing to obtain its full potential benefits, because the companies were failing to obtain deliveries of purchased materials when required for processing. There is an almost fatalistic tendency in some management teams, to accept this condition as normal and unchangeable. There are, however, other companies which have tackled the problem with persistence and courage, and have succeeded in making improvements.

Another type of control considered in the chapter is costing. The present research did not go deeply into the subject. There is an obvious need for further research in this field. This chapter only states the alternative solutions which are possible, and have been used up to now.

The chapter is based on research carried out by teams from Bradford and Salford Universities and the London Business School.

9.2. PRODUCTION CONTROL

Production control has been defined as the function of management which plans, directs, and controls the material supply and processing activities of an enterprise.

It is responsible in other words for telling the buyer when bought materials and finished parts should be delivered to the factory and for telling the managers of component processing departments and groups inside the factory, which parts should be manufactured, what quantities should be made and the delivery date or dates by which they should be completed.

9.2.(a) Levels of production control

Production control can be seen as taking place at three main levels. These are:

(1) *Programming.* At this level plans are made for the output of finished products. These plans take the form of a series of production programmes.

(2) *Ordering.* At this level plans are made for the input of purchased materials and parts from suppliers, and for the output from component processing departments and groups. The plans in this case normally take the form of 'purchase delivery schedules' or 'purchase requisitions' for the buyer, and of 'shop orders' for the processing departments.

(3) *Dispatching.* This level of planning is sometimes called machine loading, operation scheduling, or by a number of other terms special to different companies. At this level plans are made showing the sequence in which the operations required to complete parts, should be carried out on the machines in the department or group. The plans in this case, often take the form of a Gantt chart. The information from these charts is sometimes abstracted daily, to produce lists of jobs to be done, known as daily work sheets, or daily planning sheets.

Each of the above levels of production control will be considered in turn. The case of a factory machining parts for the assembly of its own products will be studied first. The case of companies which make components for sale will be examined later in the chapter.

9.2.(b) The objectives of production control

It is comparatively simple to quote the objectives which must be achieved by any successful production control system. In the case of a company making components and assembling them into finished products, for example, these objectives might be to so plan and control production that:

(1) products are available for delivery to customers when required;
(2) the random and seasonal variations in the receipt of sales orders, are 'smoothed' to produce a production programme which gives a reasonably even load on production capacity;
(3) assembly achieves the plans given in the production programme;
(4) sets of parts are available when required for assembly;
(5) materials are received from suppliers before the times when they are needed for processing.

The research has shown that very few companies in the engineering industry are succeeding in achieving these objectives. One of the main reasons for this is that few production control systems are planned or operated with these objectives in view.

One common deficiency is that sales forecasts and production programmes are seldom produced with the regularity and precision needed to plan component production and material supplies efficiently. Another deficiency in many systems, is that the objective of providing sets of parts for assembly is given a very low priority. These systems concentrate on subsidiary objectives, such as 'maintaining an economic (sic) load on the machines'. The need to balance the stocks into sets for assembly, is left to the 'progress chasers'.

Production control is fundamentally one of the simpler functions of management. It is possible to design simple production control systems which can achieve the objectives of production control with precision and regularity. The problem with most production control systems in engineering today is that they are much too complicated, because they concentrate on dispatching at the expense of progressing and ordering and because they attempt to optimise variables which can only be optimised at the expense of reliable delivery.

9.2.(c) Opposing variables

One of the facts of production which is often ignored when designing production control systems is that some factors can only be optimised at the expense of others. This is particularly true in the case of 'machine utilisation' and 'delivery to customers by promise date'.

To optimise machine utilisation, it is desirable that there should be considerable stocks of work in progress, so that there are always jobs waiting to go on the machines. It also appears to be desirable that parts with long set-ups times should be made in big batches, to reduce total setting time. Such a strategy inevitably induces a high investment in stocks, high materials obsolescence – because the parts are never in balanced product sets – and tends to generate 'shortages', because capacity needed to eliminate shortages is tied-up making parts in big batches for stock. This strategy also suffers from the deficiency that it is extremely complicated to operate. The present research has shown that most of the companies which use it are finding it impossible to make it work efficiently.

To optimise delivery to customers by delivery promise dates, on the other hand, it is essential that promise dates should be based on the production programme, that assembly should achieve the production programme and that product sets of parts should be available when programmed for assembly. The only reliable method for achieving this last requirement, is to make the components in sets as they are needed for assembly. This strategy certainly leads to improved customer service, lower stocks and reduced materials obsolescence. It is also much simpler to operate and requires less indirect labour. To make it work, however, it is essential that sales forecasting and production programming should be better than is generally the case today. There is also a need to keep production capacity and the requirements of the production programme in balance. It is desirable again, that there should be some measure of labour flexibility. Finally, this strategy emphasises the importance of choosing methods, tooling and loading schedules, which will reduce setting times.

There is no doubt that the second of these strategies is the correct one for Group Technology. What is more, because it is simpler to

Input of materials
I

Output of products
O

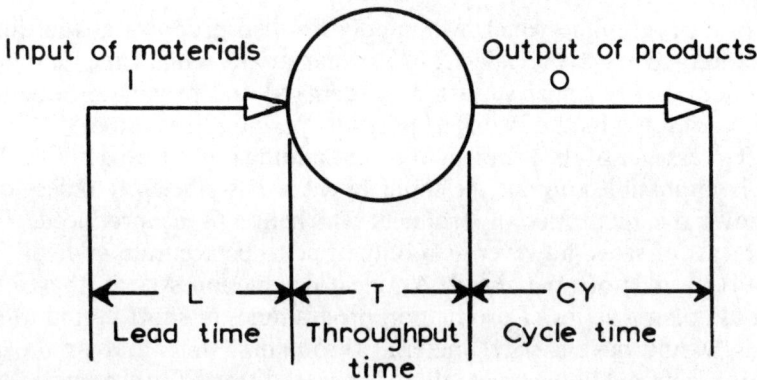

← L → |← T → |← CY →
Lead time | Throughput time | Cycle time

1. To plan 'I' one must forecast 'O'.
2. The forecast must be made for any cycle, at least (L+T) days before it starts.
3. The accuracy of 'O' depends on the time ahead for which the forecast is made.
4. 'T' (the throughput time) is a function of the complexity of the Material Flow System and of the cycle.
5. 'L' (the lead time for materials) is a function of Purchasing and Stocking policy.
6. REDUCING THE CYCLE INCREASES FLEXIBILITY.

Fig. 9.1 The essence of programming
(*Reproduced from Burbidge, J.L.*, The Introduction of Group Technology, *1975 (Heinemann, London)*)

operate and more reliable, it also tends to improve both labour productivity and overall plant utilisation, as has been shown in practice by the increased product output and increased output per employee achieved with its use.

9.3. PROGRAMMING

The first level in production control is concerned with the planning of the production programmes. These programmes should show the quantities of products to be produced in a series of equal periods, and be based on the information provided by marketing in the form of either customers orders or sales forecasts showing the products they expect to sell.

The main aim in framing the production programme should be to plan a programme which will supply finished products at the times required to meet the sales forecast and at the same time 'smooth' random and seasonal variations in demand and provide production with as even a load of work as possible, period after period.

The essence of the problem of programming is illustrated in Fig. 9.1. It is impossible to plan the input of materials efficiently unless one knows the quantities of products which are to be produced. The forecast of sales, however, can only hope to be accurate in detail for short periods of time ahead. An ideal production system, therefore, would plan a series of production programmes at short period intervals, would have a short material throughput time, and would use purchasing methods which allow short lead times for changes in the rate of purchase deliveries. Group Technology coupled with new methods of purchasing, makes this possible. It is possible therefore with Group Technology to have a flexible production system which can follow short term changes in market demand.

9.3.(a) The choice of period

The choice of programme period is mainly a function of the complexity of the product. The programming periods chosen should, ideally, be such, for Group Technology, that the same periods can be used for ordering on both assembly and the component processing groups. In practice two-week periods have been used successfully in a number of companies making fairly simple products such as pumps and valves. Four-week periods have been used successfully with more complex products such as machine tools.

The problem when choosing the programming and ordering period, is to balance the gains with a short period, such as a reduced investment in work in progress and an increase in the flexibility to follow market changes, against the losses which may be caused by an increase in the number of set-ups.

9.3.(b) Production programme term

Two main types of production programme are required:

(1) *An annual programme.* This will show the forecast production programme by periods for each product covering 12 months ahead. As each period ends during the year, one new period should be added to complete the annual production programme.

(2) *Period programme.* A new period production programme should be planned and issued at the beginning of each period. This will be based on the latest up to date forecast of the sales requirement for the appropriate period, taking into account any stocks of finished products.

The main uses of the annual programme are for financial forecasting and in production control for the control of purchase deliveries. The more accurate the annual sales forecast on which it is based, the less frequently will the annual production programme have to be revised and suppliers delivery schedules have to be changed.

The series of single period programmes, on the other hand, will be used mainly as described later to control assembly and the issue of manufacturing orders to component processing groups. It is particularly important with Group Technology that these production programmes should be based on well planned sales forecasts and should be produced systematically and at regular intervals.

9.4. ORDERING

Traditionally most ordering systems have been based on multi-cycle ordering. In other words different batch quantities and batch frequencies have been used for different components. One of the reasons for this has been the widely held belief in the principle of the 'economic (sic) batch quantity'.

9.4.(a) Multi-cycle ordering systems

One example of a multi-cycle ordering system which is still the most commonly used ordering system in industry, is 'stock control ordering'. With this method an 'economic (sic) batch quantity' is fixed for each item, and a new batch of this size is ordered every time the stock of the item drops to its own special re-order level.

Stock control ordering systems suffer a number of disadvantages which make them unsuitable for use with Group Technology.

(1) *High stocks.* They only work efficiently with large batch quantities. The stock investment, particularly in work in progress, is therefore high.

(2) *Obsolescence.* They are the main cause of materials obsolescence.

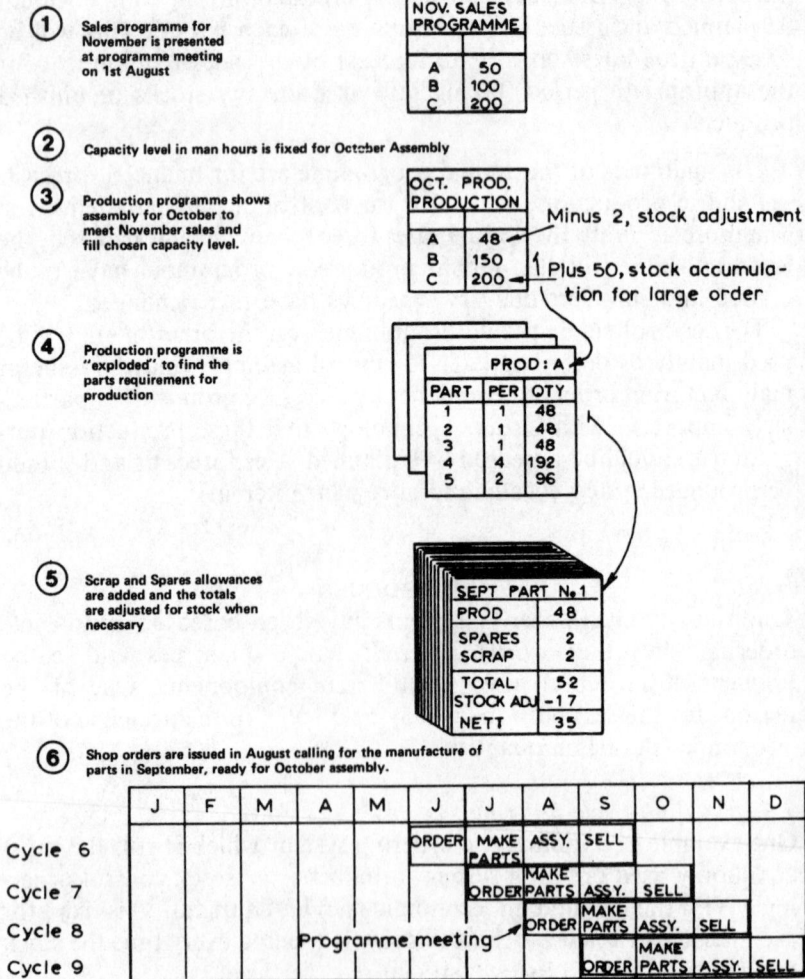

(1) Sales programme for November is presented at programme meeting on 1st August

NOV. SALES PROGRAMME

A	50
B	100
C	200

(2) Capacity level in man hours is fixed for October Assembly

(3) Production programme shows assembly for October to meet November sales and fill chosen capacity level.

OCT. PROD. PRODUCTION

A	48	Minus 2, stock adjustment
B	150	
C	200	Plus 50, stock accumulation for large order

(4) Production programme is "exploded" to find the parts requirement for production

PROD : A

PART	PER	QTY
1	1	48
2	1	48
3	1	48
4	4	192
5	2	96

(5) Scrap and Spares allowances are added and the totals are adjusted for stock when necessary

SEPT PART No 1

PROD	48
SPARES	2
SCRAP	2
TOTAL	52
STOCK ADJ	-17
NETT	35

(6) Shop orders are issued in August calling for the manufacture of parts in September, ready for October assembly.

	J	F	M	A	M	J	J	A	S	O	N	D
Cycle 6						ORDER	MAKE PARTS	ASSY.	SELL			
Cycle 7							ORDER	MAKE PARTS	ASSY.	SELL		
Cycle 8								ORDER	MAKE PARTS	ASSY.	SELL	
Cycle 9									ORDER	MAKE PARTS	ASSY.	SELL

Programme meeting

Fig. 9.2 Period Batch Control
(*Reproduced from Burbidge, J.L.,* The Introduction of Group Technology, *1975 (Heinemann, London)*)

Because parts are not made in balanced product sets, the introduction of a new design of product or a modification tends to scrap large quantities of parts.

(3) *Variable machine load.* If used to generate orders for a machine shop, basing ordering on the stock levels in a finished parts store, for example, they provide a highly variable and unpredictable load of work on the machines. With Group Technology this means that the load of work issued to a group will vary greatly from period to period.

(4) *Tooling families.* Because different parts are produced to different cycles, the different parts in any one tooling family will not all be on order at the same time. The advantages of reduced setting time and of increased capacity, are, therefore, unobtainable with a multi-cycle ordering system.

9.4.(b) Alternatives to multi-cycle ordering

One alternative to multi-cycle ordering is single cycle ordering in which all parts are ordered to the same cycle. The orders are issued to the groups at the beginning of each period and all orders must be completed by the end of the period. Research at Bradford University and the London Graduate School of Business supports this idea that a single cycle ordering system is essential for Group Technology.

Research at Salford University on the other hand, supported the idea that there should be a limited range of ordering cycles, varying between one and six different cycles, depending on the particular conditions found in a company. As will be shown later, these two ideas are not as far apart as they appear to be at first sight.

The two alternatives to traditional multi-cycle ordering proposed here are then:

(1) single cycle ordering – 'Period Batch Control';
(2) limited cycle ordering – 'Cellular Batch Quantity'.

Each of these will be described in turn.

9.5. PERIOD BATCH CONTROL

The first alternative to multi-cycle ordering systems such as those described earlier, is single cycle Period Batch Control, (PBC). The basic method in this case is illustrated in Fig. 9.2. It will be seen that the ordering of parts is based on 'explosion' from a series of period production programmes, and that the exact quantity of parts required to meet each period production programme is ordered, subject only to adjustments for additional parts for sale as spares, additions added as scrap allowances, and any necessary stock adjustment due to faulty spares and scrap forecasts in the past.

This is not a new method. It has been used for many years even with traditional forms of organisation. It is, however, much more efficient with Group Technology because the shorter throughput times make it possible to work with much shorter periods than was possible with traditional organisation.

9.5.(a) *Where PBC is used*

Period Batch Control is normally used, with only a few exceptions, to control the ordering of all made parts. One possible exception is that if, say, a four week period has been adopted, it may be decided to control the processing of some class A high cost made items to a shorter cycle in order to reduce the stock inventory. This can be done, for example, by planning the output from a heavy machine shop machining big castings, to meet the needs of a weekly assembly programme.

A second exception concerns the Pareto analysis class C or low value items. In many general engineering companies most of the class C items are purchased 'fittings'. There are some companies, however, which make large quantities of class C items. In these cases there may be advantages in reduced setting time, if the class C items are made to a longer cycle. This will increase the stock investment, but because the items have a low annual usage value, the effect on the total inventory will be small.

Consider, for example, the case of a company using PBC with a four week cycle. One group based on multi-spindle automatics, makes a large family of class C components. The method and tooling development necessary to reduce set-up times to a level where a four week cycle can be used efficiently, may take several years. The family of parts in this case could be divided into four sub-families, each of which is ordered in turn to a 16 week cycle. The division into sub-families should be based in this case, on 'tooling families', and each sub-family should be approximately equal in load. The quantities to be ordered will have to be found by 'explosion' from the annual production programme, and some buffer stock will be needed to accommodate variations between the annual programme and the series of short term period programmes.

Ideally, PBC would also be used to control purchase deliveries. In mass production the supplier can be sent a 'call-off note' at the beginning of each period, telling him the quantities he is to deliver against his supply contract in say, four weeks time. In smaller companies this is difficult to arrange. An effort should be made, however, to obtain at least the more expensive class A items by this method.

9.5.(b) Economic effects of changing to PBC

It is difficult for production engineers who have been brought up in the tradition of the 'economic batch quantity' to see that the gains when one changes from a multi-cycle ordering system to a single cycle ordering system can greatly outweigh any disadvantages. This change substitutes major reductions in the stock investment, plus major reductions in materials obsolescence, plus the advantages of an even load of work, plus major savings in set up time and increases in capacity, plus reductions in indirect labour costs due to the simplification of the control system, for rather doubtful gains in cost and output due to the use of large batches for parts with long set up times. Most production engineers only start to believe that the change is possible when they see the savings in setting time which can be achieved with a planned sequence of loading and tooling families, and begin to understand that further big reductions in setting time are also possible through technological development.

There is no doubt that an increase in sales which increases batch quantities will reduce costs. What is disputed is that one can obtain the same advantage by making some of the parts for low volume products in large batches covering long periods ahead. Such a policy induces additional costs which more than cover the gains from large batches.

9.5.(c) Sub-assembly stages

The example of PBC given in Fig. 9.2 shows the simple case of a single stage of assembly. Some products are designed so that they can be assembled in stages, producing progressively, for example, sub-sub-assemblies, sub-assemblies, assemblies and finished products. In traditionally organised industries with many assembly stages, such as the electrical switch gear industry, for example, the low orders of assembly are often made for stock. Such companies generally carry a very heavy investment in stocks and work in progress and achieve very low rates of stock turnover.

With PBC the 'explosion' to find the component requirements can be carried out in stages, progressively, to find the requirements at the different levels of assembly. Additional periods are then added to the standard ordering schedule (see Fig. 9.2, (6)) for each assembly stage.

Each period added to the standard ordering schedule, however, increases the throughput time, increases the work in progress and reduces the flexibility to follow changes in market demand. There are advantages therefore if the number of assembly stages can be reduced. This can be done, for example, by planning assembly so that assemblies are produced progressively in the same assembly group and during the same period, as main product assembly. Again it is often possible to plan assembly groups which combine several early stages in assembly. The job enlargement represented by this approach also has the advantage that it is usually welcomed by the workers.

9.5.(d) PBC and the computer

There are a number of synonyms for PBC, which have been introduced recently by computer companies. Of these the best known are 'Materials Requirement Planning' (MRP), and 'Requirement Planning Method' (RPM).

Most of the computer companies now have 'program packages' which can be used for this method of ordering. Typical of these are:

ICL – Nimms
IBM – RPG, Prince and PCS
NCR – IMMAC

All of these programs are designed to find component and material requirements by explosion and implosion from finished product production programmes. All are capable of explosion at a number of levels. They are generally designed to include the purchasing of materials as the final level. As will be seen later, there are difficulties with this last level, but the programs can be used without it. These programs are also designed to accommodate different batch frequencies for different components. Again, however, this part of the program need not be used.

As far as ordering is concerned, the introduction of the computer has been of enormous benefit. For this purpose and for the associated tasks of loading and stock records, it works with simple reliable data, much of which is fixed for relatively long periods of time. Providing that the accuracy of the data is maintained, the computer can do these jobs much more accurately and much faster than is possible manually.

9.6. LIMITED CYCLE ORDERING – THE CELLULAR BATCH QUANTITY (CBQ) METHOD

Research at Salford University supports the idea that it is essential to limit the number of ordering cycles. They recommend a variation in the number of different cycles of between one and six, depending on the particular conditions in a company.

They give as an example company A which manufactures components ranging from complex and expensive machined castings, to tiny pins manufactured on multi-spindle automatics, with the cost of components ranging from several hundred pounds to fractions of a penny. As a second example they give company B, which manufactures a range of machined parts whose costs vary between £50 and £5. The researchers at Salford suggest that company A should use a four, five, or six cycle system, whereas company B should have used a single cycle or PBC system.

9.6.(a) The CBQ method

The method is illustrated in Fig. 9.3 and summarised in the following description of the six steps required to use it.

(1) *Collect and compile the data listed in Fig. 9.3.*

(2) *Calculate the spare set up capacity of a group or machine shop.* This is the maximum number of times the key machines in a group can be set up within a given period (6 or 12 weeks) which will leave sufficient capacity to enable the required demand to be met on time. This depends not only on the machine capacity but also on the number of setters required and available.

(3) *Carry out Pareto analysis of components.* To do this it is necessary to determine the cost of the components to be manufactured and the estimated six or 12 month demand. The Annual Usage Value (AUV) or component cost times its demand is then calculated. Pareto

1　　COLLECT DATA

- List of components
- Operation times for each component on each machine.
- Set up times for each operation and hence average Set up times for each machine type
- List of machines available
- Annual demand adjusted to typical demand for each component
- Cost of each component material labour and overheads
- Calculate and list annual usage values (AUV) for each component

(AUV = component cost x annual demand for components)

2 CALCULATE SPARE SET UP CAPACITY

3　　CARRY OUT PARETO ANALYSIS OF CELL OR MACHINING GROUP

% of Annual Usage Value

Components

4 CHOOSE BATCH FREQUENCY SERIES
e.g 48,24,12,4,2,1

5　　ALLOCATE CAPACITY AND FREQUENCY

Allocate 80% of spare set up capacity divide components into cost groups and allocate a batch frequency to each group of components

% of AUV

48　24　12　4　2　1

Components

6 CALCULATE BATCH SIZES FOR EACH COMPONENT

$$\text{Batch Size} \doteq \frac{\text{Demand}}{\text{Frequency}}$$

7 MEASURE WORK IN PROGRESS LEVEL AND BALANCE SAVING IN REDUCTION IN WIP AGAINST COST OF EMPLOYING ADDITIONAL SETTERS AND ADJUST NUMBERS OF SETTERS ALLOCATED TO CELL

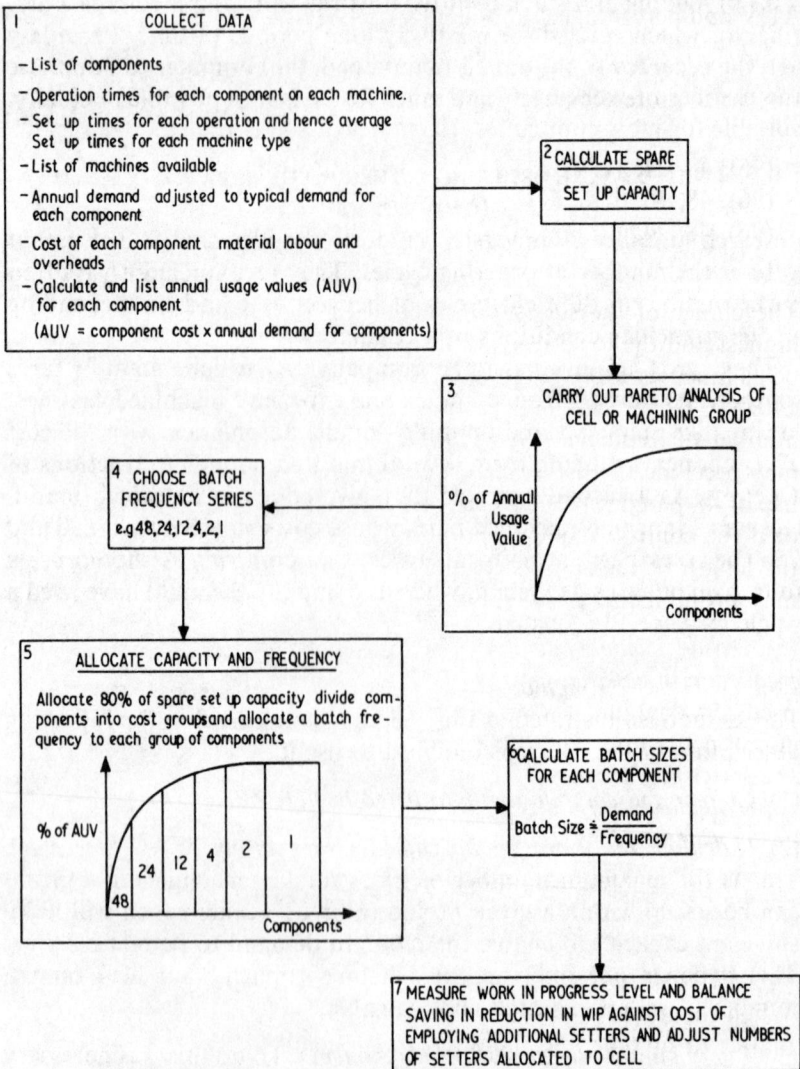

Fig. 9.3 Cellular Batch Quantity method

analysis is then carried out to find the relationship between percentage AUV and percentage component variety.

(4) *Choose a batch frequency series.* The series chosen will depend on the number of working weeks in a year. Generally 48 is found to be suitable for most companies. Examples of some frequency series are:

(96), 48, 24, 12, 4, 2, 1,	(64), 32, 16, 8, 4, 2,
(96), 48, 16, 8, 4, 2, 1,	50, 25, 5, 1,
(96), 48, 24, 12, 6, 3, 1,	50, 10, 5, 1,
	52, 26, 13, 1,

The choice of the series will depend on the difference between the highest and the lowest AUV. In some companies this difference may be as much as £100 000 to £1m. In others it may be only £100 000 to £10 000. Generally the greater the difference the greater will be the number of frequencies required. There are certain limitations which may preclude the use of the extreme values such as 48 and one. The use of 48 batches per year may be left out if capacity is very limited or if the control period is more than one week. At the other extreme the length of the run if a component is made in one batch per year may prohibit the use of this frequency.

(5) *Divide components into cost groups and allocate a batch frequency to each cost group.* If a series such as 48, 24, 12, 4, 2, 1, is used, it contains six numbers so it is necessary to divide the components into six groups and allocate a frequency number from the series to each as shown in Fig. 9.3.

(6) *Calculate batch size.* The batch size of any component can be easily computed by dividing the demand by the chosen frequency.

$$CBQ = \frac{Demand}{Frequency}$$

9.6.(b) Operating and maintaining the CBQ System
When the initial analysis has been carried out it is a relatively simple matter to up date and maintain the system. In most companies new products are continuously being developed and old products are being modified or phased out. Fluctuations in demand may change product mix and it is necessary to up date the information which is

used to calculate batch sizes. Minor changes can be accommodated easily but it will be necessary to recalculate batch sizes once a year or once every two years. The major part of the effort is expended in collecting data but if the data is continuously updated this becomes a relatively simple exercise. One of the simplest methods of collecting and storing data is on punched cards which are useful both for feeding data into a computer and can also be used as a manual system. A card can be designed to record all the relevant information about each component.

Component number	– B204597
Drawing number	– 1234
Cell number	– 3
Cost of components (each)	– £0.50
Estimated annual demand	– 10 000
AUV	– £5000
Batch frequency	– 2
Batch size	– 5000

The cards for each group can be stored together in the order of their AUV.

If the demand for a component increases in such a way that its new AUV falls in a different frequency group it is simply a question of moving the card forward or backward into the relevant frequency groups. If changes in demand affect several components in such a way that all the spare set up capacity is used up, it may be necessary to re-adjust the frequencies of other components which fall on the boundary line between two cost groups.

9.6.(c) Comparison of CBQ and PBC methods

The fundamental difference between the two methods is that the CBQ method fixes a batch quantity for each component. The PBC method recalculates the batch quantities each period by explosion from a series of period programmes. Apart from this difference the two methods have much in common.

Where there is a limited range of AUVs, the CBQ method adopts one standard batch frequency. In this case the CBQ and PBC methods can be very similar.

When there is a wider range of AUVs and the CBQ method chooses two or three different frequencies, the two methods can still be similar. For example, with the PBC method, a standard period of four weeks might be chosen for the short term programmes, giving a standard frequency of 12 batches per annum, but some exceptions might be accepted. Some high value class A items might for example be made in 48 batches per year and some low value class C items might be made in four batches per year. The CBQ method with three batch frequencies would in this case find a similar solution.

It is only when the CBQ method selects four, five, or six batch frequencies that it differs radically from the PBC method.

9.7. DISPATCHING, MACHINE LOADING, OR OPERATION SCHEDULING

Dispatching is the third level of production control. At this level orders will have been issued to the group and the next job is to plan the sequence in which the different operations required to make the parts are started on the different machines in the group.

One interesting finding of the present research is that the computer may not be the right tool for operation scheduling. In several cases where computer scheduling is used, the schedules produced by the computer bear little relation to the way in which the work is actually loaded on the machines in the machine shop. In fact the only cases where computer scheduling appears to work, are those where the computer schedule is used only as a guide to the foreman and he can vary this schedule where he thinks fit. Under these circumstances one can logically question if the computer schedule is really necessary at all.

This finding is not a criticism of computers or a criticism of the men who make the scheduling programs. The real problem is that it is impossible or at least uneconomical to feed into the computer the masses of up to date and rapidly changing data which are necessary if a near optimum schedule is to be found. The technological data required for scheduling can be provided fairly accurately. Other types of necessary information, such as the serviceability of machines and tools, the effects of different loading sequences on setting time, absenteeism, the special characteristics of different machines of the same type, the special skills of different operators, the preference of different operators for working certain machines, jobs with tight or loose rates, late arrival of materials, sub-standard materials, and many other similar items of information, are very difficult to provide for the computer and to keep up to date. These latter types of information are common knowledge in a group. It appears likely, therefore, that the foremen in the groups are in a better position to make reasonable schedules than any centralised scheduling service, with or without a computer.

As the foreman in each group is the only one likely to have sufficient information to produce a reliable schedule, it seems probable that the best policy for 'dispatching' is to train the foreman in the elements of the subject and to provide him with any special information, such as lists of parts made from the same types of material, the part numbers of parts in the same tooling families, order priority lists, and so on, so that he can make his own schedule. It seems possible that it might be more appropriate to use the computer to provide the foreman with these types of information, than to use it to tell him exactly what to do. One advantage of this method is that the foreman can react quickly when disasters such as machine breakdowns and tooling breakdowns make it necessary to change the schedule. In one company making complex products in relatively large groups, a work scheduler has been attached to each group, to help the foreman with this work.

There is one other reason why the use of the computer for work scheduling should be questioned. There is evidence that the opportunity to plan and control their own work is highly valued by foremen and workers and makes an important contribution to their job satisfaction.

9.8. LOADING AND SMOOTHING

The receipt of sales orders for products is likely to vary from period to period due to random and seasonal variations. A sales forecast is therefore generally unsuitable as a basis for the control of production. It is generally necessary to translate the sales forecast into a production programme which smooths out the variations in load imposed by the sales forecast and provides a relatively even load of work, period after period, for the factory.

The best method of smoothing for assembled products is by varying the stock of finished products. It is necessary that the capacity of the factory to produce finished products should be maintained as near as possible in line with the trend of market demand. It is also necessary that the capacity at different stages of processing should be in balance. For example, if there is sufficient capacity to assemble a given programme in one period, there should be sufficient capacity in the machine shop to machine the parts for these products in one period, and sufficient capacity in the foundry to make the castings required for this period output. If this is achieved, smoothing can be effected by making the main products for stock when demand is low and using this finished product stock to supplement production when the demand is high.

In the past, with traditional systems, many companies have attempted to smooth production and to reduce delivery times by adjusting component ordering and holding heavy stocks of made materials and machined parts. With the very much shorter through-put times achievable with Group Technology, this represents an unnecessary tieing up of capital, unnecessary additional costs, and an unnecessary risk of materials obsolescence.

With single cycle ordering it is very simple to calculate the load in machine hours on each machine in a group, which is imposed by any given production programme. Each product will impose a certain load in machine hours on each machine and the total load imposed by the product will then depend on the numbers of products in the programme. Ideally the load used should be the nett load. In other words it should not include setting time. Setting time varies enormously with loading sequence and the use of standard setting times per component is, therefore, unreliable. The nett load must be compared with the nett capacity of the machines in the groups, or in other words with the available hours minus forecast total setting time and idle time.

The research done by Bradford University included a number of studies of the actual variations in load, being experienced in groups. Examples of the 'load profiles' produced, are shown in Figs 9.4 and 9.5. Figure 9.4 gives two examples of the way in which the load was distributed between the machines in a group, in a typical period. Figure 9.4 (a) shows a fairly even balance of load between the machines, indicating a reasonable design of group. It illustrates the way in which the proportion of setting time varies with different types of machine. It also shows that production control was overloading

Machine
Hours

Total load

Week beginning 21.4.75

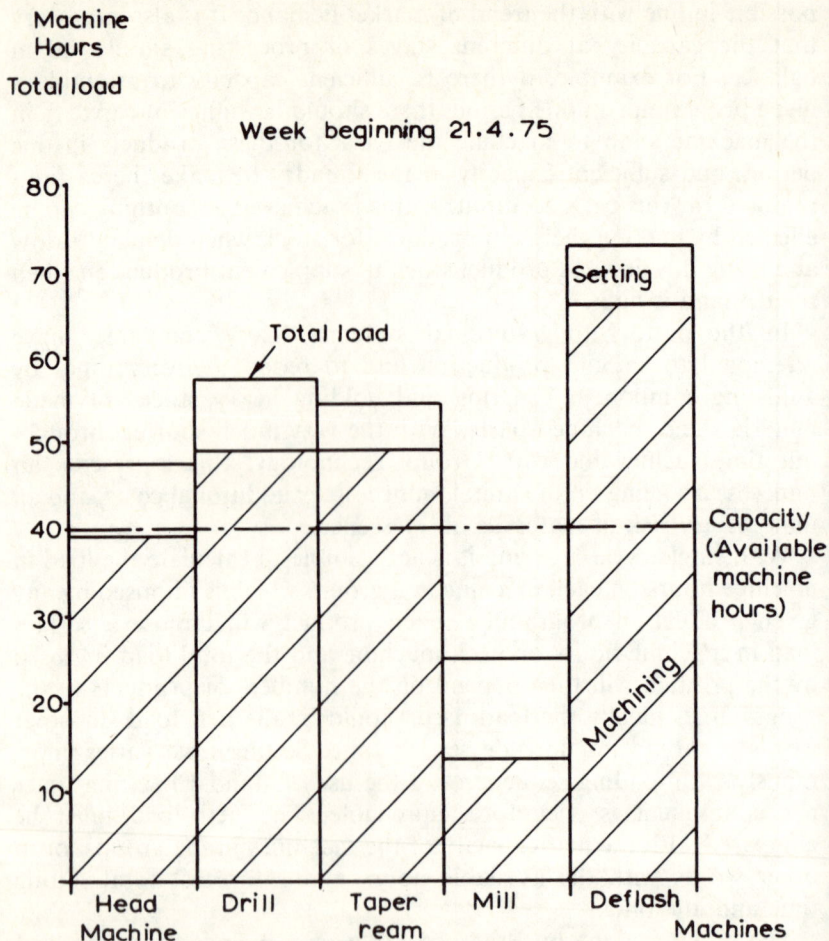

(a) Group (F), Q Ltd

Fig. 9.4 Load distribution of groups

the group beyond their capacity, due in this case to inadequacies in the ordering system.

Figure 9.4 (b) shows a more complex group. It shows a typical distribution of load between machines. Two of the underloaded machines are special to this one group. The change to groups is not therefore responsible for their lack of load balance.

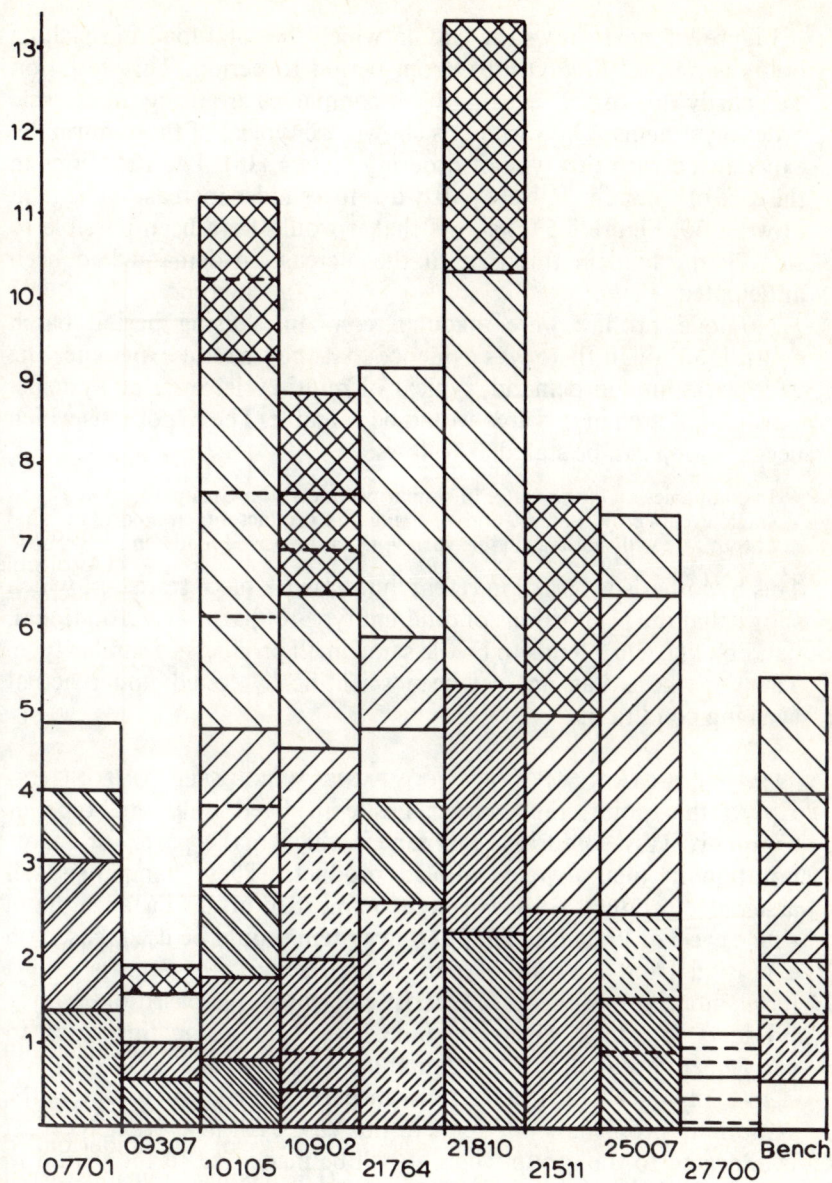

(b) *Group (4), R Ltd*

Fig. 9.4 Load distribution of groups

Figure 9.5 next shows the way in which the total load in machine hours varied in three groups, from period to period. This variation was partly due to the fact that both companies are using multi-cycle ordering systems. The variations shown are typical of those normally experienced with this type of ordering system (16). The variations in the case of company P, were partly due to a sudden increase in demand in week 50. Figure 9.5 (c) shows that it would have been possible to smooth the load in this case, if the increase in demand had been anticipated.

No load profiles were made in companies using period batch control, although there was evidence that they do not experience the wide variations in demand, typical of multi-cycle ordering systems. Further research in this area would be valuable. The hypothesis which needs testing can be stated as follows.

> In companies making a range of similar products with Group Technology and single cycle Period Batch Control, changes in product mix at a constant total output rate will cause negligible variations in the period loads on the groups.

This hypothesis is based on research into work piece statistics, which shows that the proportions of different types of parts – e.g. rotational, flat, box like, etc – tends to be the same in all products of similar type. The hypothesis has not yet been scientifically tested under actual working conditions.

9.9. PRODUCTION CONTROL IN COMPANIES WHICH SELL COMPONENTS

Up to this point, the chapter has considered only one type of company. This comprises companies which make parts for incorporation in their own assembled products. These companies sell assembled products. It has been recommended in this instance that a single cycle or limited cycle ordering system should be used, based on explosion from a series of period production programmes. Short term smoothing in this case is most efficient if it is based on making products for stock when demand is low, and using this stock to supplement production when demand is high.

In companies which make finished parts for sale to customers the smoothing problem is different. In this case each new order received is allocated to the group where it will be made and to the period in which it must be completed if it is to be delivered on time. This is continued until there is no more capacity in a group for a period. At

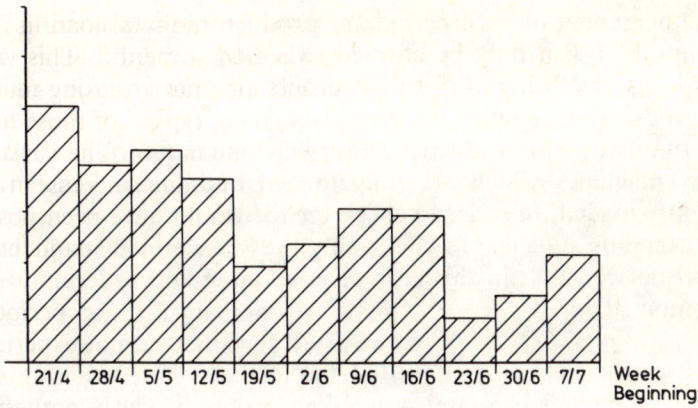

(a) *Group (A), Q Ltd*

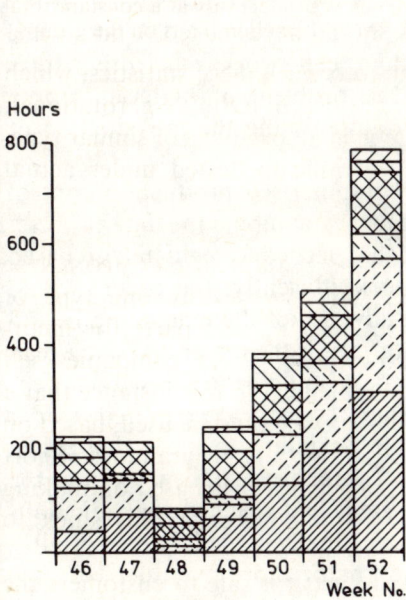

(b) *Group (9), P Ltd*

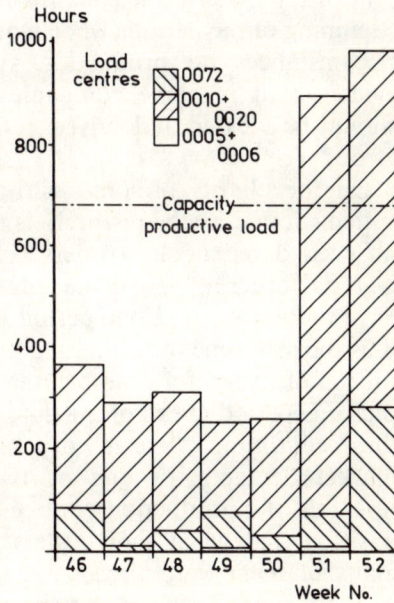

(c) *Group (3), P Ltd*

Fig. 9.5 Variations in period load

the beginning of each period the production programme is 'sealed'. Smoothing can only be effected, when the period load is less than capacity, by bringing forward orders previously allocated to later periods.

In this type of production, it is likely that there will be a number of key machines which are fully loaded and others which are only lightly loaded. In order to check the loads allocated to each period it is generally sufficient to check only the loads on the key machines. It will be seen that in this case programming and ordering are partly combined.

9.10. PURCHASING

The research has found a number of companies where the main difficulties in production control are caused by unreliable deliveries of materials and finished parts from suppliers. In one company 79 per cent of the materials needed for processing were not available at the beginning of the periods when work should have started. Under these circumstances no production system can work efficiently. It is evident that in some companies the problem of material supply should be tackled and solved before attempting the introduction of Group Technology.

One peculiarity of some of the computerised production control systems found by the research, is that they combine the data processing needed to issue shop orders with that needed to regulate purchased material ordering. To make this work efficiently, the orders would have to be processed one period in advance of the machining period plus the lead time for obtaining materials. If the materials have a lead time of one year, for example, the order should be processed one year plus one period ahead of the date when the parts are required. Under these conditions there is no possible flexibility to meet changes in market demand. In one or two companies studied during the research, the production control system had reached the ridiculous position that when shop orders were finally issued to the groups, many of them were already six months or more overdue. No production control system can hope to maintain respect under such conditions.

In most companies the obvious solution is to separate purchasing from shop ordering. One solution is to base purchase deliveries on an

Part No.	Description	1	2	3	4	5	6	7	8	9	10	11	12	13	Orders O/S
	Period No.														
13876		(22)	(22)	(25)	(25)	25	30	25	25	22	22	22	20	20	45
13902		(22)	(22)	(25)	(25)	(25)	(30)	(25)	(25)	(22)	22	22	20	20	Nil
13990		(44)	(44)	(50)	(50)	50	60	50	50	44	44	44	40	40	350
13992		(22)	22	25	25	25	30	25	25	22	22	22	20	20	150
14010		(50)	(50)	(50)	(50)	(50)	55	55	55	50	50	50	50	50	75
14018		(50)	(50)	50	50	50	55	55	55	50	50	50	50	50	400
14072		(50)	(50)	(50)	50	50	55	55	55	50	50				100
14076		(50)	50	50	50	50	55	55	55	50	50				50
14078		50	50	50	50	50	55	55	55	50	50				250
14111		(100)	100	100	100	100	110	110	110	100	100				500

Fig. 9.6 Explosion from annual programme. Used to control deliveries of purchased castings

Key
(XX) — *stocks of finished parts*
XX — *stocks of bought castings*
XX — *no stocks*

explosion from the annual production programme showing period requirements, and to allow some buffer stock to accommodate variations between the annual programme and the series of period programmes used to control machining.

Figure 9.6 illustrates the explosion from an annual programme, showing the stocks of finished parts and bought castings available, and indicates the period when more castings are likely to be needed. The purchasing objectives in this case can be stated in such terms as: 'At the beginning of each period there must be at least one period's supply of each part in stock and not more than three periods' supply'. The purchasing manager is then left to negotiate delivery schedules with the suppliers which will maintain stocks within these limits. It is desirable that he should be responsible for his own stores and stock records.

In large companies one possibility is to divide the purchasing department into groups, each dealing with a different class of material. In one company, for example, it is planned to form procurement groups, each responsible for its own stock records, delivery records, order issue and progressing and each dealing with a different class of materials. These will be:

Group 1, responsible for the purchase of all raw materials.
Group 2, responsible for the purchase of all class A purchased parts.
Group 3, responsible for the purchase of class B and C purchased parts.
Group 4, responsible for the purchase of supplies and miscellaneous purchases.

It is essential for efficient purchasing that the annual sales forecast should be as accurate as possible. Both it and the annual production programme should be revised where necessary, but too frequent revision of the annual programme and of the schedules given to suppliers, is likely to frustrate the suppliers and reduce their belief in the need to supply according to the schedule.

9.11. PROGRESSING
One advantage of Group Technology is that because each part is always made in its own group, there is never any doubt where it can be found. The progressing of shop orders is ideally best left to the groups themselves. They should be given the responsibility for completing all the work allocated to them for a period, by the end of that period. As each part is finished and delivered to stores or to assembly, they will cross it off their list. Towards the end of the period they will be able to see very clearly which items are still outstanding and need progressing. The groups should be encouraged to report disasters which are likely to delay the completion of their batch of work early. Operation progressing should be unnecessary.

The progressing of purchase deliveries is again best left to the purchasing department. They will know what quantities of purchased items they are expected to deliver at the beginning of the next period and will have all the information necessary to judge where special hastening action is required. Their problem is really one of choosing good suppliers and keeping them up to the mark. The method of 'supplier assessment' is one of the tools which they can use to achieve reliable deliveries.

9.12. COSTING SYSTEMS
Most of this chapter has been concerned with production control. It was mentioned, however, at the beginning of the chapter, that

there are other supporting systems which may need changing to suit Group Technology. One of these is the costing system.

In most traditional costing systems the cost centres are machine tools, machining sections, and departments. With the change to Group Technology the obvious cost centres are the groups. Some modification of the costing records will be essential in most cases.

It is an unfortunate fact that little has been published by accountants up to the present time, to describe the changes necessary in costing to suit Group Technology. There appear to be two schools of thought on this matter.

(1) *The group as a cost centre.* There are those who believe that the group should be the smallest cost centre and that cost can be controlled by recording the groups total performance.

(2) *Operation records.* Secondly there are those who believe that full control of cost is not possible without details of the time spent on each machine inside the group on each operation.

Those who believe that groups should be made as independent as possible and should be fully responsible for their own performance and those who see as a main advantage of Group Technology the possibility to reduce paper work and indirect labour, are likely to favour the first solution. They believe that it is still possible to control costs by measuring the period output from the group in terms of standard cost and comparing this with the expenditures incurred in running the group.

The present research noted the problem in several companies but did not provide solutions. All that can be stated here is that there is an obvious need for further research in this area.

9.13. CONCLUSIONS

The present research has clearly demonstrated that for complete success with Group Technology, it is necessary in many companies to change the production control methods.

The most appropriate method of production control for Group Technology is Period Batch Control. This method means in most companies that they will be making parts in smaller batches. The fact that all parts are ordered together in product sets at regular period intervals means, however, that setting times can be greatly reduced by using a planned sequence of loading on the machines with the highest set up times. In several cases this has led to a reduction in setting time which was not only sufficient to compensate for the smaller batches and therefore more frequent set ups, but also gave a nett increase in capacity.

The research has also indicated that a main difficulty with production control in many companies, is due to a failure in their materials supply system. There appear to be advantages if the data processing for the control of purchasing is separated from the data processing for the control of shop orders.

Again, the research has indicated that except in the simplest industries, the computer may not be the right tool for dispatching, or operation scheduling.

The computer is ideal for such jobs in production control as explosion, implosion, load calculations and stock records. In these cases relatively fixed data is used and the computer can do the job more accurately and quickly than is possible manually. It is also socially desirable in these cases, moreover, because it eliminates clerical drudgery.

The data required for scheduling, on the other hand, is not fixed. Much of it varies from hour to hour. The computer can not be used efficiently in such circumstances. Its use for this purpose is also socially undesirable, because it removes from the floor of the shop the highly valued possibility of planning their own work.

Finally some changes may be necessary in other supporting systems. They may be necessary, for example, in costing.

CHAPTER 10

THE FUTURE POTENTIAL
OF GROUP TECHNOLOGY

10.1. INTRODUCTION

The researches covered by this chapter were carried out by a team from the London Business School. They included a number of studies which attempted to evaluate what was already being done in Group Technology, and one which tried to assess the potential market for groups in the mechanical engineering industry.

The chapter starts by looking again at the limitations which may affect the use of groups. It attempts to answer the questions: where can groups be used?, where are groups now used?, and finally, where could groups be used? These studies give an indication of the possible future potential for the use of Group Technology.

10.2. WHERE CAN GROUPS NOT BE USED?

The engineering industry is mainly concerned with the manufacture of assembled machines. It is therefore mainly concerned with component processing and assembly and has little to do with technological systems of the process industry type.

10.2.(a) Limitations to the use of groups

The size of industrial unit appears to offer no restriction to the use of groups. As will be shown later, very small companies exhibit most of the desirable characteristics of groups and are in effect already groups. At the other end of the scale very large companies with 5000 employees or more have installed groups.

The sophistication of the technology used again appears to offer no limitation to the use of groups. Successful applications of Group Technology include, at one extreme, companies making parts for aero engines and very sophisticated airborne radar equipment. At the other extreme there are companies making very simple machines and components, which have also successfully introduced Group Technology.

Those writers who have seen limits to the utility of Group Technology, have been mainly concerned about batch quantities. They believe that Group Technology is ideal for batch production, but that it is unsuitable for the very small quantities of each component required in some forms of jobbing production and also that it is unsuitable for the very large quantities made by continuous flow lines in mass production. This point will now be considered first in relation to component processing and then in relation to assembly.

10.2.(b) Component processing
Group Technology in Britain has up to now been mainly concerned with the batch production of components in machine shops. Many examples now exist, varying in the size of the company, the complexity of the product, the type of market supplied and in the degree of mechanisation. There seems to be very little doubt now that Group Technology can be used for any type of component processing in batches, which can be found in the engineering industry.

Looking next at jobbing production and particularly that part of the jobbing industry which makes small numbers of special parts to order, examples can be found of companies which have used groups, although they are few in number. These jobbing groups exhibit all the seven desirable characteristics of groups which were defined in Chapter 2, they are therefore groups within the working definition of a group adopted for this book.

These groups in jobbing factories also achieve most, if not all, of the benefits claimed for Group Technology in batch production. For example, because each component is made and completed in one group, throughput times are short, leading to short delivery times. The investment in stock and work in progress tends to be lower due to the shorter throughput times and to the centralisation of responsibility for all processes on each component in the groups. Production control tends to be simpler and requires less paper work, materials handling costs are reduced. Savings in setting time on the other hand are of only minor importance as most machines in this type of industry are general purpose machines which can be quickly set up for different types of part.

Although the total of savings in a jobbing factory are likely to be less than they would be in a batch production factory, they are still generally worthwhile. It appears that Group Technology is not only possible in jobbing machines shops but is also both economically and socially desirable.

Finally in the case of component processing, it is necessary to consider the instance of the mass production of very large quantities of components on maching lines. It has been suggested that there can be no possible advantage in 'changing to groups' in these cases. It is almost certainly true that no change is required. Almost all of these lines are, however, relatively small. They generally employ 15 or fewer workers per line. Looking back at the working definition and the desirable characteristics of groups described in Chapter 2, it will be realised that these small lines are already groups, within the terms of the definition. There is certainly little to be gained by changing them except in details of organisation.

To summarize, there appear to be no limitations to the use of Group Technology in the case of component processing.

10.2.(c) Assembly

The use of groups in assembly in Britain, has been mainly restricted to jobbing production and small quantity batch production. In these cases it is common to find groups of workers who specialize in the complete assembly of particular type of product or sub-assembly. Many of these groups again, although not generally known by that name, fit the working definition of a group and exhibit most of a groups desirable characteristics.

When the quantities of products to be produced is larger, even if they are made in batches, it is common to make them on an assembly line. When these lines employ a relatively small number of workers they, like the machining lines described in the previous section, fit the working definition of a group and exhibit many of their desirable characteristics. In the classification of group organisation (19) they are 'Z' Groups, or groups internally organised for progressive flow.

Finally, if the quantities to be assembled are very large, as in the assembly of motor vehicles and consumer durables, the usual method of assembly has been to adopt a conveyorized assembly line. These lines are not groups because they generally employ a large number of workers.

The conveyorised assembly line has been seen for many years as the ultimate in economic production. They have the disadvantage, however, that they provide very simple tasks for the workers on the line which are monotonous, boring and tiring. This has led to high absenteeism, high rates of labour turnover and industrial unrest in many of the companies which have used conveyorised lines. In some countries in Eurpe today the conveyorised assembly line is no longer an economic proposition. It may seem to be the best solution on paper, but it is uneconomic if one cannot get workers to accept work on the line.

Groups in mass production assembly were therefore introduced mainly for economic reasons which were caused by social problems. When they were introduced it was widely expected that the change would increase costs, but that it was necessary to meet changes in workers needs and expectations and would in the long term prove to be the only way to keep companies in operation. In practice most of the companies which have changed from conveyorised lines to group assembly have found that they have achieved both economic and social advantages.

The immediate economic benefits obtained by changing to groups in assembly are not of the same high order as those which can be achieved in component processing. They are nevertheless positive and support the contention that there are no limitations to the use of groups in assembly.

10.3. WHERE ARE GROUPS NOW USED?

Many groups already exist in industry. These will be considered under three headings. First, long standing groups in factories; second, groups in the form of very small companies, and third the new wave of groups based on Group Technology.

10.3.(a) Long standing groups

Many companies in the British engineering industry already contain at least one or two groups. They may not be called groups at present but they fit the working definition of groups and they exhibit most of their desirable characteristics.

Typical examples are the tool rooms found in small and medium sized companies. These tool rooms make completed tools (products), they are provided with all the facilities needed to make them, these facilities are laid out in one area and so on. They are in effect groups. Other examples include a crank shaft group and a cam shaft group in a factory making diesel engines and a special group making biscuit cutters in a factory making food machinery. In addition many of the component processing lines found in industry are, as explained above, already groups in everything but name, and many assembly sections are in fact groups within the working definition of a group.

Groups are, in fact, not a completely new idea. They have been used where it seemed possible throughout industrial history. What is new about Group Technology is that it attempts to use groups in a much wider variety of circumstances.

10.3.(b) Very small companies
One of the studies undertaken by the London Business School in the present research, was a study of the 3000 companies employing 15 or fewer employees in the British engineering industry. This study first looked at the desirable characteristics of groups. It was found that these small companies exhibit most of the desirable characteristics of groups and are therefore in effect groups.

The number, of 3000 companies with 15 employees or fewer, was found by extrapolation from figures in the '*Digest of Statistical Information—1971*'. This shows that there are some 5940 enterprises in the industry which employ less than 25 people. A rough extrapolation found that there are somewhere in excess of 3000 companies which employ between five and 15 people. These companies comprise 35 per cent of all companies in the British mechanical engineering industry.

The study then went on to look at the productivity of these small factories. It was found that these small units do not, on the surface, appear to be as efficient production units as the larger, traditionally organised companies. Again quoting from the digest of statistical information, for the group of companies employing less than 25 people, the nett output per head was £1351 per annum, compared with £1485 per annum for companies employing 400 to 1000 people, and with the industry average of £1370 per annum.

All the literature published by the companies using groups has stressed the financial benefits of Group Technology. The groups are claimed to give greater efficiency including increased output per man. Yet here are 3000 single group companies which exhibit all the desirable characteristics of groups and yet have an output per man below the industrial average. These efficiency figures, of output per man, however, have two components. One is the efficiency of the production unit to manufacture products cheaply and quickly, and the other is the efficiency of the commercial side of the business to obtain work at a realistic price. It has long been the contention of organisations representing small businesses that they are highly efficient as production units but totally unrealistic commercially. At present no data is available to show whether this contention is correct, but a sensitivity analysis of the figures available is useful.

In the mechanical engineering industry approximately 30 per cent of output cost is labour cost. The average labour cost is £1480 per annum. Therefore the average cost of output per man is £4930 per annum. Thus adding the nett output per man to the cost of the output per man (£4930 + £1370), should give the company quoted prices of output per man of £6300 per annum. The nett output per man in the small companies is £19 per annum below average and £134 below the maximum. In order to achieve the average the companies would only need to increase prices by 19 divided by 6300 or 0·3 per cent. In order to achieve the maximum they would need to increase their prices by 134 over 6300, or 2·1 per cent. For a section of industry which is poor at quoting realistic prices, these figures indicate a very efficient production organisation which can achieve 99·7 per cent of the average industry performance with below average prices.

The small single group companies have shown that even with a hazardous and notoriously poor commercial system, the production unit can still obtain output figures within 2 per cent of the maximum achieved in the whole industry. Larger companies achieve highly efficient commercial systems but seem to loose control of the production unit. A combination of the two, the commercial performance of the large unit and the production performance of the small, makes a formidable concept.

10.3.(c) Recent Group Technology groups
Finally, the remaining groups which can be found in British industry today are those which have been introduced since the advent of the new ideas of Group Technology.

There are approximately 19 086 different establishments shown in the statistics for the British engineering industry. Approximately 70 of these companies have made some move towards Group Technology. Most of these only have a small number of groups working. There are about 26 which have completed a major change to Group Technology. This means that something less than 0·4 per cent of the companies in the British engineering industry, excluding the very small companies which are already groups, have made some attempt to change to Group Technology. Of these, however, only approximately 26 have succeeded in making a major company wide change.

10.4. WHERE COULD GROUPS BE USED?

It has been shown that there is no absolute limitation to the use of groups in component processing or assembly. They can be and have been used both in Britain and elsewhere **(19)** to make almost every conceivable type of engineering product. It has also been shown that at present in Britain less than 0·4 per cent of the establishments in the engineering industry have as yet made an attempt to introduce Group Technology. It is now necessary to study where groups could be used.

10.4.(a) Industry wide
There are those who say that Group Technology can be used in any type of company. They see Group Technology as a fundamental change in organisation which bases the division into organisational units on the product or its components, instead of on the process. They point out that in nearly every case the change to groups has given some economic benefits and also had social advantages. They believe that Group Technology represents a new basic concept of organisation which should replace an older and, until recently, universally accepted concept, based on the principles of the division of labour, process specialisation and functional organisation.

There is obviously some truth in this idea. It should be remembered, however, that the economic gains from the introduction of Group Technology depend in part on the technological system concerned and on the type of organisation which existed before.

For example, consider a company employing, say, 400 workers, engaged in jobbing production using very heavy machines, in which the workers have a high level of skill and most of the technological details are planned on the floor of the shop. In such a case the managing director might legitimately say, 'I could change to Group Technoloy, but my estimates show that the economic benefits I could achieve would be relatively small and the social advantages would also be small, because our workers already have a high degree of participation. On the other hand the cost of moving our heavy machines would be very high and I do not believe that the rate of return on this investment would be worthwhile'.

It will be seen that although Group Technology is possible and will normally give some economic and social advantages, there may be limiting cases where only minor benefits will be achieved. The problem at the moment, however, is that too many companies are making this kind of excuse, without any real justification.

10.4.(b) The cream of the market
Another study by the London Business School looked at this problem of assessing the probable market for Group Technology from another point of view. This assessment was based on finding the companies most likely to gain by introducing Group Technology. It was in effect trying to find those companies which had the potential to achieve results similar to those achieved by Serck Audco Ltd of, for example, a 60 per cent reduction in work in progress, and a 50 per cent increase in output per employee.

The study was based on published statistics of which the main sources were the *Business Monitor—Census* and the *Annual Abstract of Statistics*. The first of these gives a breakdown of the mechanical engineering industry into 17 sub-sections and the second translates this information into manufacturing environment terms. The investigators had personal knowledge of applications of cellular manufacture in 15 out of the 17 sectors.

The basic approach to assessing the market for groups was, firstly, to consider each of the 17 sectors of the mechanical engineering industry separately and secondly, to use a carefully developed framework to assess the potential for Group Technology in each of these sectors. The resultant individual values were then combined to provide a total potential 'market' for groups.

The framework was devised by considering an average or typical company in each sector and was based on three factors or dimensions; the size of the manufacturing unit, the typical batch quantity, and the extent of assembly required by the product. The last two dimensions were derived from an appreciation of the type of product involved. By considering the complexity of the product and the normal proportion of bought-out components, it was possible to assess the proportion of operators in assembly. Having split manufacturing into assembly and component processing, the potential for the use of Group Technology was then assessed.

In component processing, industries which normally work with either very small batch quantities (jobbing) and industries which work with very large batches (mass production) were excluded. Typically, batch quantities of between 10 and 1000 were considered to indicate a high potential for Group Technology.

In the case of assembly the limits imposed were that if the work could be brought to the man and was concerned with medium sized batches, Group Technology was feasible. If long runs were normal, then flow line manufacture was more appropriate. All companies with 25 employees or less were excluded from the study.

The estimate of the number of groups (cells) which could be formed in the mechanical engineering industry is given in Fig. 10.1. The total estimate was 15 000 groups with 185 000 workers employed within them. This represents approximately 20 per cent of the total work force of the industry. If the individual sub-sections are considered, the areas of greatest potential can be listed as follows.

(1) Metal working machine tools.
(2) Pumps, valves and compressors.
(3) Miscellaneous non-electrical machinery.
(4) Office machinery.

These four sub-sections represent 72 per cent of the estimated total potential and indicate the specific areas in which management should certainly be seeking to introduce Group Technology. It is interesting to note that the first three of these four sections do not display powerful economies of scale. The companies are generally relatively small. In office machinery the main potential for groups was seen in assembly.

Sector	Construction and earth	Misc. (non-elecal) machinery	Mech. handling equipment	Food & drink process equip.	Ordnance & small arms	Space heating, ventilating & a/cing equip.
Total No. of Cells	673	1536	686	85	23	768
Total No. of Cell Operators	8090	18 433	8240	1027	276	9225
Method of Sub-Dividing the sector for analysis	Large batch & rapid throughput	Large & complicated m/cs 25-400 over 400 people	Standard Products	Companies of 25-400 people	Companies of 100-400 people	Domestic market
No. of Assy Cells	285	Cells 232 / Cells 344	225	Cells 37		116
No. of Assy Cell Operators	3420	Ops 2787 / Ops 4125	2700			1390
No. of Comp Mfg. Cells	183	Cells 387 / Cells 573	336		Cells 23	460
No. of Comp Mfg. Cell Operators	2200	Ops 4646 / Ops 6875	4040	Ops 448	Ops 276	5530
(sub-division)	Small slow batches	Light mafg.	Specials	Companies over 400 people		Industrial market
No. of Assy Cells	80		50	Cells 48		38
No. of Assy Cell Operators	970		600			460
No. of Comp Mfg. Cells	125		75			154
No. of Comp Mfg. Cell Operators	1500		900	Ops 579		1845
Total Assy Cells	365	Cells 1536	275	Cells 85		154
Total No. of Assy Cell Operators	4390	Ops 18 433	3300		23	1850
Total Comp Mfg. Cells	308		411			614
Total Comp Mfg. Cell Operators	3700		4940	Ops 1027	276	7375

Fig. 10.1 (a) Potential for groups in UK mechanical engineering

Sector	Agric. m/cs	Metal Wking m/c Tools	Pumps valves & compressors	Ind. engines	Mining m/cs	Printing book bind.	Refrig. m/cs	Textile m/cs	Office m/cs	Ind. (incl. process) Plant & Steelwork	General mech. eng.
Total No. of Cells	538	4415	3800	316	144	300	175	708	1183	No potential	No potential
Total No. of Cell Operators	6464	53 000	45 700	3830	1739	3600	2119	8511	14 224		
Method of Sub-dividing the sector for analysis	Companies of 25-400 people	Companies of 25-400 people		25-750 people	Companies of 25-400 people	Companies of 25-400 people		25-500 people	25-750 people		
No. of Assy Cells	113	833		17	19		63	39	280		
No. of Assy Cell Operators	1357	10 000		210	229		756	474	3380		
No. of Comp Mfg. Cells	113	1916		26	28		31	158	27		
No. of Comp Mfg. Cell Operators	1357	23 000		320	343		378	1898	330		
	Companies over 400 people	Companies over 400 people		Over 750 people	Companies over 400 people	Companies over 400 people		Over 500 people	Over 750 people		
No. of Assy Cells	156	500		110	39		54		800		
No. of Assy Cell Operators	1875	6000		1320	467		657		9600		
No. of Comp Mfg. Cells	156	1166		165	58		27	511	76		
No. of Comp Mfg. Cell Operators	1875	14 000		1980	700		328	6139	914		
Total Assy Cells	269	1335		127	58		117	39	1080		
Total No. of Assy Cell Operators	3232	16 000		1530	696		1413	474	12 980		
Total Comp Mfg. Cells	269	3082		191	86	300	58	669	103		
Total Comp Mfg. Cell Operators	3232	37 000		2300	1043	3600	706	8037	1244		

Fig. 10.1 (b) Potential for groups in UK mechanical engineering

10.5. CONCLUSIONS

The studies reported in this chapter have shown that there are substantial opportunities for the introduction of Group Technology in the British mechanical engineering industry.

If the 3000 small establishments, which employ 15 or fewer workers and already have most of the desirable characteristics of groups, are excluded, less than 0·4 per cent of the establishments in the industry are at present using Group Technology.

Estimates of the potential for Group Technology vary between the whole industry and 20 per cent of the total work force of the mechanical engineering industry. The 20 per cent figure is undoubtedly very conservative. The method of assessment used excluded parts of the industry where groups are already being used efficiently. Again, it only looked at the 'mechanical engineering industry' category of companies, and excluded the electrical, electronic, automotive, aerospace and shipbuilding components of the engineering industry, all of which have successful applications of Group Technology. It has the advantage, however, that it points clearly to sectors of industry where maximum economic advantage can be obtained very easily.

CHAPTER 11

CONCLUSIONS

This chapter summarises the conclusions reached by the research described in this book. The research has been concerned mainly with the engineering industry, including both mechanical and electrical engineering categories. It has dealt mainly with component processing in these industries. Its findings are therefore mainly applicable to companies with machine shops, press shops, sheet metal shops, forges and other similar departments.

11.1. THE MANUFACTURING SCENE

Many recent studies have shown that the British engineering industry shows up badly in comparison with its competitors in productivity, in rates of stock turnover, in exports and, particularly, in its apparent inability to deliver goods to its customers by promised dates.

Economists and others have suggested a number of reasons for this state of affairs. Most of them have seen the cause of these problems in the environment in which industry works. The present research has indicated that a much more probable cause of the industries difficulties lies in its failure in the management of production.

The present research has indicated that the main problems in the engineering industry can be attributed to deficiencies in the way in which the material supply and processing activities are directed and controlled in our factories. It shows that in many factories in this industry at present, this part of management is out of control. In other words less than 90 per cent of the plans made for production are actually achieved on time.

This failure in production is seen as the result of two main factors. Firstly, it has been caused by an excessive belief in process specialisation and functional management. On the line organisation side process specialisation produces very complicated material flow systems, through which it is extremely difficult to control the flow of materials. On the staff management side, functional management means that most organisational units of managers are composed of specialists and few have the authority necessary to complete major tasks. The main problem of management today is therefore to co-ordinate the work of many different specialists.

The second cause of our present difficulties is excessive growth in the size of our production units. At the factory floor level this has led to a reduction in the content of the tasks which workers do, making their jobs boring and unsatisfying. On the management side this growth has led to an increase in the number of levels in the organis-ation and has been followed, in Britain particularly, by a tendency to centralise most of the policy making in large headquarter units, manned by teams of specialists. This has had the effect of leaving the plant managers and their subordinate line managers with no real authority to make decisions. These managers are frustrated by the fact that they are not able to make necessary decisions and to take the necessary action to overcome the day-to-day problems which are the reality of life in a factory.

There is evidence that while 'bigness' may give financial and market-ing strength, it is disastrous for production. It is noticeable that the few highly efficient large companies in Britain are those which have pursued a policy of establishing small production units based on products and have delegated a high level of authority and responsi-bility to their factory managers and inside these factories to the line managers running departments and smaller groups.

There is evidence that Britain has ample reserves of capacity, that exports could be increased if delivery performance could improve and that our prices are already competitive and could be made more so if the difficulties in production could be overcome.

It has also been shown that many of Britain's difficulties in industrial relations spring from the way in which production is organised. The social system is changing and workers expectations of what they want from their jobs are higher than they were some years ago. They expect, in particular, today, some say in the way in which their own work is planned and controlled. This is difficult to arrange efficiently with present methods of organisation, but simple and effective with Group Technology.

What is needed today is for production systems to be designed as simple systems composed of largely independent units, to which the authority and responsibility for production performance can be clearly allocated in simple terms. If this is done and simple objectives are established covering major requirements, the production managers and the men on the floor of the shop are more competent to make the system work and to overcome the random mischances of life in a factory, than is possible with any combination of formal centrally controlled systems run by independent staff specialists. These types of organisational system are those produced by Group Technology, which presents at the moment the most promising way in which present difficulties could be overcome.

The studies made have indicated that the initiative for change must come from management if it is to be effective. There is very little which government can do which would be effective in this field. The best contribution it could make would be to provide stability of the total economy and consistancy of industrial policy.

11.2. DESIRABLE CHARACTERISTICS OF GROUPS

The research has indicated that Group Technology is fundamentally a new philosophy of organisation which seeks to replace the present long established philosophy based on the division of labour and Taylor's functional management.

The traditional form of organisation uses organisational units which are based on processes at shop floor level and on functions in management. No organisational unit in this case completes products or major management tasks.

The new methods of organisation known as Group Technology, form organisational units at assembly level which are based on the product, or at an earlier stage in production on completed components, or again at an even earlier stage on completed material items such as castings and forgings. In the case of staff managers, again, they form organisational units which, in terms of our present specialisations, are multi-disciplinary groups with all the necessary skills to complete major management tasks.

The basic building block of Group Technology organisation is the group. The desirable characteristics of groups are as follows.

(1) *Team.* There should be a set of workers special to the group.

(2) *Facilities.* There should be a set of machines and equipment which is special to the group.

(3) *Group Layout.* These machines and equipment should be laid out in one special area reserved for the group.

(4) *Products.* The group should complete its own special set of products.

(5) *Target.* Tasks and output targets should be given to the group as a whole and not separately to the different individuals in the groups.

(6) *Size.* The number of workers in each group should be small enough to obtain social cohesion.

Any organisational unit which exhibits these characteristics is a group.

Groups with these characteristics have been designed and introduced by both production engineers and behavioural scientists. The groups formed by engineers were designed mainly with operational and economic objectives. The groups formed by behavioural scientists were usually formed with social objectives in view. The fact that both have succeeded in forming groups which met their objectives, indicates that Group Technology can solve many of the economic and social problems of industry, both at the same time.

With the sole exception of size, these same characteristics are seen as the desirable characteristics of larger organisational units such as departments and divisions. Some authorities have suggested in fact that the best way of introducing Group Technology is to start from the top and work down.

11.3. EXPERIENCE

The research has found sixty seven examples in Britain of companies which have introduced Group Technology. Twenty six of these have achieved a complete change in at least one of their departments.

Combining British experience with that in other countries, it is difficult to find any type of product which has not already been produced in groups. It is difficult also to find any type of production system for which there is not already a successful application of

Group Technology. Groups have been used with highly sophisticated technology in, for example, the aeronautical industry and in fully automated factories. At the other extreme they have been used in factories making simple products with old and unsophisticated plant. Size, again, does not appear to offer any restriction to the use of Group Technology. Applications can be found in a small jobbing machine shop employing 44 workers and at the other extreme in a number of companies with 5000 workers or more. There do not appear to be any restrictions to the use of this new type of organisation.

The research has indicated that the companies which have been most successful with Group Technology have been those which have designed groups, which have most clearly followed the desirable characteristics of groups listed above. Not surprisingly, as the companies which have introduced groups so far were the pioneers, some of them have failed to achieve all the desirable characteristics. It might be said that there are still two main jobs to the done in Group Technology. The first is to persuade more companies to make the change and the second is to develop some of the existing applications to obtain their full possible advantages.

11.4. BENEFITS OF GROUP TECHNOLOGY

The research has indicated that all the companies which have introduced Group Technology are claiming some benefits. These vary in degree from the major benefits obtained by the ten or twelve best examples, to smaller gains obtained by companies which have been less successful in implementation. These benefits are in three main categories. Firstly, there are operational benefits, secondly economic benefits, and finally social benefits.

The operational benefits might be summarised as a considerable simplification of the material flow system. This, at least in the best installations, has brought production back under control and made it more effective and easier to manage.

This simplification of the production system has also induced major economic benefits. Examination of the published claims for successful applications gave the following results.

Gains	Average (per cent)	Maximum (per cent)
Reduction in work in progress	62	85
Reduction in total stocks	42	44
Reduction in throughput time	70	97
Increase in output per employee	33	50
Reduction in overdue orders	82	—

There is little doubt now that Group Technology can produce substantial economic benefits which could not be obtained in similar circumstances with our traditonal forms of organisation.

The studies into the social effects of Group Technology financed by the Science Research Council are being undertaken by a team at Birmingham University. This team started work later than the others and has not yet completed its research. Its preliminary studies have indicated however, that the change to Group Technology does eliminate many of the features of the traditional form of organisation which have been criticised by sociologists, phsychologists and by the workers themselves.

11.5. DESIGNING GROUPS

The research has studied the methods used by different companies to plan the introduction of Group Technology. It suggests, firstly, that although some companies have made major changes in their products and in production methods before introducing Group Technology, it is possible and probably better to make the change with the existing products, machines and methods first and then to carry on with other types of development afterwards.

It also studied the ways in which responsibility and authority have been allocated to individuals for the task of planning and introducing Group Technology. It noted that the most successful applications have all been led by a senior director in the company, and suggests that without this leadership from the top, it is difficult to introduce Group Technology efficiently, because all levels and functions in the organisation are likely to be affected.

Under this director there should be an individual, or a small team, responsible for detailed planning and for the co-ordination of implementation. The question whether to use outside help was examined. It was suggested that the ideal method for the planning and implementation of Group Technology requires a company team, plus the assistance of outside experts with a wider knowledge of other applications.

After the policy has been decided a systematic approach to the design of Group Technology is recommended. This follows the stage of defining objectives, designing the production system specification, any necessary departmental re-organisation, finding the groups, development of the groups in relation to group manning, group management, plant layout, tool storage, quality control and materials handling. Finally, in order to obtain a fully integrated and stable system it is necessary to re-plan some of the related systems such as product design, sales and marketing, production control, costing, personnel management and production planning to suit the new organisation.

The research studied the methods used to plan the division into groups and families. It recommends a comprehensive analysis, based on all the parts, to obtain a total plan for the change, using the technique of Production Flow Analysis.

11.6. IMPLEMENTATION AND EARLY OPERATION

The research looked closely at the problems of implementation and early operation. By 'implementation' is meant the actual physical changes in layout and the changes in procedures which are made when Group Technology is introduced. The term 'early operation' covers the period between starting work in the groups and achieving a steady state.

The research has clearly indicated the types of difficulty which may be experienced in the implementation of Group Technology and in its early operation. It is obvious that constant vigilance is needed if the full benefits of the new system are to be obtained.

In the early days of Group Technology, planning was restricted almost entirely to the planning of the groups on the floor of the workshop. Today it is realised that changes in other supporting systems are also essential for complete success.

It is submitted that it is possible to plan and control the introduction of Group Technology systematically with considerable precision. It is also possible to make reliable estimates of cost and economic benefits, once the preliminary planning has been done.

The main problems with early operations have been due to deficiencies in the information system, problems with production control, problems with purchasing, problems with work scheduling and progressing and problems with the methods of supervision. In addition, not enough attention has been given in many of the applications to the training of those involved.

11.7. PERFORMANCE MEASUREMENT

Studies have also been made of the information needed by production managers to do their jobs efficiently. It is submitted that present methods of measuring performance are not sufficient to give the production manager the information he needs to control production efficiently.

The types of information he needs in addition to the usual financial information, include such measures as:

(1) overdue deliveries to customers;
(2) purchase orders overdue;
(3) quality rejects on purchased orders;
(4) shop orders not completed by due date;
(5) quality rejects on shop orders;
(6) labour turnover;
(7) absenteeism.

These and other similar measures of performance are known as 'physical performance measures'. These types of information are usually available somewhere in most companies, but today they are seldom recorded systematically and presented at regular intervals for the guidance of the production manager. They are essential, however, if he is to carry out his job efficiently.

These types of measure are also required with Group Technology to give the groups the information they need about their performance in meeting their targets and objectives. Without efficiently organised feedback to give the groups enough information to monitor their own performance there is no possibility of achievement motivation.

The research provides a framework for the development of performance measurement in the future. It provides an important missing element in the information services provided for management.

11.8. CONTROL

The research has clearly indicated that in many companies most of the difficulties being experienced with Group Technology can be attributed directly to production control. Traditional methods of production control do not work efficiently with groups and cause such problems as a widely fluctuating load of work from period to period.

Two of the research teams support the thesis that the best production control system for use with Group Technology, is single-cycle Period Batch Control (or MRP). This is an 'explosion method' which finds the quantities of components and materials to be ordered by calculation from the programme for the final assembly of products, or in the case of component products, from the programme for deliveries to customers. The team from Salford University proposed an alternative method based on a limited number of batch frequencies.

The research has also indicated that some companies are having difficulties with Group Technology due to irregular deliveries of purchased supplies. Late deliveries from suppliers is due, in many companies, not solely to the deficiencies of the suppliers. Failure to plan production efficiently and know what is needed in sufficient time is at least a contributory cause in many cases.

The research has also indicated that computer scheduling to plan the sequence of loading work on the machines in a group, is not the ideal method. There appear to be both economic and social advantages if this work is left to the groups themselves.

11.9. THE FUTURE POTENTIAL OF GROUP TECHNOLOGY

The research has indicated at one extreme that it is possible to use Group Technology in every kind of factory in the engineering industry. At the other extreme an attempt was made to assess those types of company in the mechanical engineering industry which could hope to achieve the major economic benefits described earlier. This study did not consider the electrical or electronics industries. A very conservative estimate found that at least 20 per cent of the mechanical engineering industry could achieve major benefits.

11.10. CONCLUSION

Group Technology offers the best hope for a rapid improvement in efficiency in the British engineering industry. A great deal of experience has been gained on how to plan and introduce the necessary changes. The main remaining and unsolved problem is how to persuade the directors and senior managers in the industry to take the necessary steps to introduce it.

REFERENCES

(1)   *Anon.*, Computer aided shop scheduling for the small firm. BPICS Bristol conference, November 1974.
(2)   BOOZ Allen report to the Department of Trade and Industry. May 1973.
(3)   1975 Conference on North Sea Oil 'UK deliveries average 15 months late'.
(4)   VICE, A, 'The Weinstock Yardstick of Efficiency', *The Times*, 29 November 1968.
(5)   SPIEGELBERG, R, *The City*, 1973, p. 82 (Quartet Books, London).
(6)   Department of Industry Survey. 1975. (Internal).
(7)   REDDAWAY, 'An analysis of Takeover', *Lloyds Bank Review*, April 1972 (London).
(8)   BOLTON, J E, *Small Firms*, 1972. (HMSO, London)
(9)   NEDO Study. 'Small companies contribute more to exports.'
(10) NORMAN and BAHIRI, *Productivity Measurement and Incentives*, 1972 (Butterworth, London).
(11) SWAN and WADDINGTON, 'Productivity measurement and in the machine tool industry, *Prod. Eng*, 1974, March/April.
(12) MADDOCK, I, 'End of the glamorous adventure', *New Scientist*, 13 February 1974.
(13) SCHONFIELD, A, *British Economic Policy Since the War*, 1958 (Penguin, London).
(14) JEWKES, J, *The New Ordeal by Planning*, 1968 (Macmillan, London).
(15) MITROFANOV, *Scientific Principles of Group Technology*, 1966. (English Translation, National Library for Science and Technology, London).
(16) BURBIDGE, J L, 1975. *The Introduction of Group Technology*, 1975 (Heinemann, London).
(17) FONTIJNE, J A, 'A third or middle way to change organisations', Proceedings of seminar on the effects of group Production methods on the humanisation of work, 1976. Turin International Centre. Awaiting publication.
(18) RICE, A K, *Productivity and Social organisation*, 1958 (Tavistock, London).
(19) BURBIDGE, J L, 'The effects of group production methods on the humanisation of work, Final report. Turin International Centre.
(20) RANSON, G and TOMS, *Production Planning and Control*, 1966 (NEDO/HMSO, London).
(21) ROETHLISBERGER, F J and DICKSON, W J, *Management and the Worker*, 1932 (Harvard University Press, Cambridge, USA).
(22) DURIE, F R E, 'A survey of Group Technology and its potential for user application in the UK', *Prod. Engr*, February 1970.
(23) KNIGHT, W A, Economic Benefits of Group Technology, IPE Conference, November 1972.
(24) LAWSON and PUTMAN, 'Group Technology challenge to job shop management', *Metalwkg Econ.*, June 1971.
(25) MARKELY, J J, 'The cell system of manufacture – a case study', *Machinery Prod. Engng*, 23 August 1972.
(26) PERRINS, H R, 'Group Technology applications in a large company', IPE Conference, November 1972.
(27) WHITFIELD, W T, 'Group Technology for one-off non-repetitive manufacture', *Prod. Engr*, March 1973.
(28) WILLIAMSON, D T N, Reshaping Batch manufacture EDC/*Mech. Eng*, 1972

(29) Anon., 'Group Technology applied to the production of hot water mixing valve components', *Machinery Prod. Eng*, 1st April 1970.
(30) Anon., 'How Alfred Herbert avoided the long drop', *Busin. Admin*, October 1973.
(31) BRITTEN, S, *The Treasury under the Tories*, 1951–64 (Penguin, London).
(32) FLANDERS, A, *The Fawley Productivity Agreements*, 1964 (Faber, London).
(33) NEW, C C, *Requirement planning*, 1973 (Gower Press).
(34) BURBIDGE, J L, 'Production Flow Analysis on the computer', *Prod. Engr*, 1974.
(35) BURBIDGE, J L, 'A manual method of Production Flow Analysis', *Prod. Engr*, October 1977.
(36) PURCHECK, G K F, 'Combinational analysis in planning for cellular manufacture', Proceedings 9th CIRP International seminar on Manufacturing Systems, 1977. Cranfield Institute of Technology, Bedford.
(37) EDWARDS; G A B, 'The management problems of introducing Group Technology', 1969. Proceedings International Seminar, Turin International Centre.

APPENDIX A   PEOPLE INVOLVED IN THE RESEARCH

The following contributed to the research on which this book reports.

(*a*) *Birmingham University*

(1) Mrs M Fazakerley.
(2) G Burrell.
(3) Mrs M Hale.

(*b*) *University of Bradford* (*and UMIST*)

This part of the research started at UMIST Manchester University and was completed at Bradford University. Some of the people involved transferred with the project.

(1) G A B Edwards (UMIST/Bradford).
(2) I F K El Essawy (UMIST).
(3) I Evans (UMIST).
(4) A Thorogate (UMIST).
(5) I Barber (UMIST/Bradford).
(6) C Bradbury (UMIST/Bradford).
(7) D Russell (UMIST/Bradford).
(8) W Sutherland (UMIST/Bradford).
(9) W S H Taylor (UMIST/Bradford).

(*c*) *London Graduate School of Business Studies*

(1) Professor T A J Nicholson.
(2) R Adams.
(3) P D Hall
(4) C C New
(5) R D Pullen
(6) A M Stanton.

(*d*) *Salford University*

(1) Professor A W J Chisholm.
(2) B Fogg.
(3) A J Norton.
(4) C V Nagarkar.
(5) R N Bagrodia.
(6) R Russel.
(7) S N J Mawjr.

APPENDIX B   LIST OF RESEARCH PAPERS

The report given in this book was based mainly on the following research papers.

(a) *Birmingham University*

(1) *Anon.*, Setting the manufacturing scene.
(2) *Anon.*, Cells in the context of user firms.
(3) *Anon.*, Some claimed benefits.
(4) *Anon.*, Control.
(5) *Anon.*, Designing cells.

Note. The research at Birmingham University started later than the rest and is still continuing.

(b) *Bradford University*

(1) *Anon.*, Manuacturing and production.
(2) *Anon.*, Cells in the context of user firms.
(3) *Anon.*, Cell design.
(4) *Anon.*, Implementation and early operation of the cell system.
(5) *Anon.*, Control.
(6) *Anon.*, The significance of load profiles and sequencing in small production groups.
(7) *Anon.*, Trantor Ltd.

Note. A much larger number of separate reports, was condensed by Bradford University to produce the above consolidated reports (1–5).

(c) *London Graduate School of Business Studies*

(1) Pullen, R D, The nature of cellular manufacture.
(2) Stanton, A M, Directory of cellular applications in UK.
(3) New, C C, MRP+GT, a new strategy.
(4) Pullen, R D and Hall, P D, The definition and characteristics of manufacturing cells.
(5) Pullen, R D, A summary of cellular manufacturing cells.
(6) Hall, P D, Company characteristics.
(7) Nicholson, P A J, Midland components Star Line tools.
(8) Adams, R, An appraisal of a cell.
(9) Hall, P D, 3000 cells in the mechanical engineering industry.
(10) Pullen, R D and Hall, P D, Performance measurement and cells.
(11) Stanton, A M, Pullen, R D and Hall, P D, Estimation of the market for cells in the mechanical engineering industry.
(12) Hall, P D, Summary of gains published by users of cellular production.
(13) Stanton, A M, Stated reasons for and against the implementation of cellular manufacture.
(14) *Anon.*, Maxi reports.
(15) *Anon.*, Mini reports.
(16) Nicholson, T A J, The challenge for production management in the UK.
(17) Hall, P D, The cash flow problems of introducing the cell system of production.

(*d*) *Salford University*

(1) *Anon.*, The design of a cell system.
(2) *Anon.*, Implementation and early operation.
(3) Nagarkar, C V, Batch size and the cell system.
(4) *Anon.*, Report describing the cell system in the sheet metal shop of Platt International (September 1974).
(5) *Anon.*, Platt Saco Lowell, sheet metal shop, interim report, Cell 4 (1975).
(6) *Anon.*, Platt Saco Lowell, sheet metal shop, interim report, Cell 7 (1975).
(7) *Anon.*, Results of preliminary study into the operation of cellular manufacture at Chubb Ltd.
(8) *Anon.*, Chubb Ltd. Progress report (1974).
(9) Bagrodia, R, Dissertation. The effectiveness of production control in a cellular manufacturing system.
(10) *Anon.*, Report describing the current operations of Baxi heating (1974).
(11) *Anon.*, Investigation into Universal group tooling at Baxi heating.
(12) *Anon.*, Report describing the current operations at Oweco Ltd (1973).
(13) *Anon.*, Initial assessment of the manufacturing problems at Oweco (1973).
(14) *Anon.*, Progress report. Cellular manufacturing feasibility study. Oweco (1973).
(15) *Anon.*, Pilot study describing the formation of Wellhead cells at Oweco.
(16) *Anon.*, Garrick Engineering. Report on cell design.
(17) Norton, A J, What does the cell system mean in practice.
(18) Norton, A J and Fogg, B, Small company management and the cell system.
(19) Norton, A J, The cell system and non-machining processes.
(20) Norton, A J, Tiffin, Fogg, B and Chisholm, A W J, Costing as an aid to manufacturing engineering.
(21) Norton, A J, Report on visit to Holland (Foker).
(22) Norton, A J and Fogg, B, Report on visit to Europe. (France, West Germany and Holland.)
(23) Hunt, R, The effects of organisational change on job satisfaction.
(24) Yogaratnum, S, Design of a manufacturing system for the production of simple machine tools.
(25) Genendra, A N, A sub-optimal schedule for cellular manufacturing systems.
(26) Avunduk, K H, Sub-optimal sequencing in cellular manufacturing systems.
(27) *Anon.*, Universal group tooling and group press-working in Russian industry.
(28) *Anon.*, Report on the availability of Universal tooling in the UK.

APPENDIX C   LIST OF UK ENGINEERING APPLICATIONS OF GROUP
TECHNOLOGY

Key

(1) Company name.
(2) Address.
(3) Products made.
(4) Process: A=assembly, M=machine, S=sheet metal, P=press, W=welding, O=office.
(5) Progress: PI=Partly implemented. IC=Implementation complete.
(6) Number of groups.
(7) Per cent of workers in groups.
(8) Workers per group.
(9) Pilot=did the company start with a pilot group?
(10) Size: Number of employees in company (direct operatives).
(11) Corp: Is the production unit the only one in the company (1), or is it one of several in a large company? (C).
(12) Technology: H=high precision, M=medium precision, L=low precision.
(13) Years since first cell introduced.

1. (1) *Baker Perkins Ltd.*
   (2) Westwood Works, Westfield Road, Peterborough, PE3 6IA.
   (3) Biscuit, chemical, foundry and printing machines.

(4)	(5)	(6)	(7)	(8)	(9)	(10)	(11)	(12)	(13)
Process	Progress	Number	per cent	Workers per group	Pilot	Size	Corp.	Tech-nology	Years
M	PI	2	10	12	Yes	900	1	M	5

2. (1) *Baxi Ltd*
   (2) Bamber Bridge, Near Preston, Lancashire.
   (3) Gas fires.

(4)	(5)	(5)	(7)	(8)	(9)	(10)	(11)	(12)	(13)
Process	Progress	Number	per cent	Workers per group	Pilot	Size	Corp.	Tech-nology	Years
S	IC	4	100	8	No		C	M	

3. (1) *Bentley Engineering Group Ltd*
   (2) New Bridge Street, Leicester, LE2 7JB.
   (3) Knitting machines.

(4)	(5)	(6)	(7)	(8)	(9)	(10)	(11)	(12)	(13)
Process	Progress	Number	per cent	Workers per group	Pilot	Size	Corp.	Tech-nology	Years
M	IC	6					C	M	

4. (1) *Binks Bullows Ltd*
   (2) Pelsall Road, Brownhills, Walsall, WS8 7HW.
   (3) Paint spraying equipment.

(4)	(5)	(6)	(7)	(8)	(9)	(10)	(11)	(12)	(13)
Process	Progress	Number	per cent	Workers per group	Pilot	Size	Corp.	Tech-nology	Years
M	PI	2	40	9	No		C		2

5. (1) *BIP Engineering Ltd*
   (2) Strestley Works, Sutton Coalfield, Warwickshire.
   (3) Hydraulic presses and plastics machinery.

(4)	(5)	(6)	(7)	(8)	(9)	(10)	(11)	(12)	(13)
Process	Progress	Number	per cent	Workers per group	Pilot	Size	Corp.	Tech-nology	Years
M.A.	PI	5		6	No	68		M	

6. (1) *BOC (Welding Products) Ltd*
   (2) West Industrial Estate, Senbigh Road, Bletchley, Buckinghamshire.
   (3) Welding torch nozzles.

(4)	(5)	(6)	(7)	(8)	(9)	(10)	(11)	(12)	(13)
Process	Progress	Number	per cent	Workers per group	Pilot	Size	Corp.	Tech-nology	Years
M	IC	134	100	9·5	No	513	C	M	5

7. (1) *BOC* (*Welding Products*) *Ltd*
   (2) Angel Road, Upper Edmonton, London, N18.
   (3) Gas welding nozzles.

(4)	(5)	(6)	(7)	(8)	(9)	(10)	(11)	(12)	(13)
Process	Progress	Number	per cent	Workers per group	Pilot	Size	Corp.	Tech-nology	Years
M	IC	6	100		No		C	M	2

8. (1) *BOC Ltd*
   (2) Skelmersdale.
   (3) Gas torches.

(4)	(5)	(6)	(7)	(8)	(9)	(10)	(11)	(12)	(13)
Process	Progress	Number	per cent	Workers per group	Pilot	Size	Corp.	Tech-nology	Years
M	IC				No		C	M	1

9. (1) *Bryce Berger Ltd*
   (2) Gloucester Trading Estate, Hucclescote, Gloucestershire.
   (3) Fuel injection, eq: diesel starters and governors.

(4)	(5)	(6)	(7)	(8)	(9)	(10)	(11)	(12)	(13)
Process	Progress	Number	per cent	Workers per group	Pilot	Size	Corp.	Tech-nology	Years
M	PI						C	H	

10. (1) *Chamberlain Industries Ltd*
    (2) Staffa Works, Argall Avenue, Leyton, London, E10.
    (3) Hydraulic motors, Power packs, rams, tube bending equipment.

(4)	(5)	(6)	(7)	(8)	(9)	(10)	(11)	(12)	(13)
Process	Progress	Number	per cent	Workers per group	Pilot	Size	Corp.	Tech-nology	Years
M	PI	2		6	No	100			1

11. (1) *Chubb Lock and Safe Co. Ltd*
    (2) Tottenham Street, London, W1P 0AA.
    (3) Safes.

(4)	(5)	(6)	(7)	(8)	(9)	(10)	(11)	(12)	(13)
Process	Progress	Number	per cent	Workers per group	Pilot	Size	Corp.	Tech-nology	Years
M.A.	IC	15	100	17	No		C	M	4

12. (1) *Chubb Lock and Safe Co. Ltd*
    (2) Wednesfield Road, Wolverhampton, WV10 0EW, Staffordshire.
    (3) Locks.

(4)	(5)	(6)	(7)	(8)	(9)	(10)	(11)	(12)	(13)
Process	Progress	Number	per cent	Workers per group	Pilot	Size	Corp.	Tech-nolgoy	Years
M.A.	IC					264	C	M	

13. (1) *Colchester Lathe Co. Ltd*
    (2) Hythe, Station Road, Colchester, Essex.
    (3) Lathes.

(4)	(5)	(6)	(7)	(8)	(9)	(10)	(11)	(12)	(13)
Process	Progress	Number	per cent	per group	Pilot	Size	Corp.	Tech-nology	Years
M	PI	4	16	8	No	250	C	H	10

14. (1) *Compair Construction and Mining Ltd*
    (2) Camborne, Cornwall.
    (3) Pneumatic tools.

(4)	(5)	(6)	(7)	(8)	(9)	(10)	(11)	(12)	(13)
Process	Progress	Number	per cent	Workers per group	Pilot	Size	Corp.	Tech-nology	Years
M	PI	6			No			M	2

15. (1) *Crabtree Vickers Ltd*
    (2) South Accommodation Road, Hunslet, Lee, LS11 5TN.
    (3)

(4)	(5)	(6)	(7)	(8)	(9)	(10)	(11)	(12)	(13)
Process	Progress	Number	per cent	Workers per group	Pilot	Size	Corp.	Tech-nology	Years
M	PI	2		9	Yes	208	C	M	

16. (1) *Crane Ltd*
    (2) Stockport, Cheshire.
    (3) Valves.

(4)	(5)	(5)	(7)	(8)	(9)	(10)	(11)	(12)	(13)
Process	Progress	Number	per cent	Workers per group	Pilot	Size	Corp.	Tech-nology	Years
M	PI				Yes			M	

17. (1) *Deloro Stellite UK Ltd*
    (2) Stratton St. Margaret, Near Swindon, Wiltshire.
    (3) Stellite alloy.

(4)	(5)	(6)	(7)	(8)	(9)	(10)	(11)	(12)	(13)
Process	Progress	Number	per cent	Workers per group	Pilot	Size	Corp.	Tech-nology	Years
M	PI	1	2	3	Yes			M	1

18. (1) *Dewrance Dresser Ltd*
    (2) Trevithick Works, East Gillibrands Estate, Skelmersdale, Lancashire.
    (3) High pressure control and isolation valves.

(4)	(5)	(6)	(7)	(8)	(9)	(10)	(11)	(12)	(13)
Process	Progress	Number	per cent	Workers per group	Pilot	Size	Corp.	Tech-nology	Years
M	PI	1	1·5	3	Yes	160		M	0·5

19. (1) *Dormer Drills (Sheffield) Ltd*
    (2) Summerfield Street, Sheffield, S11 8HL, Yorkshire.
    (3) High speed steel twist drills.

(4)	(5)	(6)	(7)	(8)	(9)	(10)	(11)	(12)	(13)
Process	Progress	Number	per cent	Workers per group	Pilot	Size	Corp.	Tech- nology	Years
M	PI	2		15	No	31	1	M	2

20. (1) *Ferodo Ltd*
    (2) Chapel-en-le-Firth, Stockport, SK12 6JP, Cheshire.
    (3) Friction linings.

(4)	(5)	(6)	(7)	(8)	(9)	(10)	(11)	(12)	(13)
Process	Progress	Number	per cent	Workers per group	Pilot	Size	Corp.	Tech- nology	Years
M	IC	6		13	No	112	C	M	5

21. (1) *Ferranti Ltd*
    (2) Ferry Road, Edinburgh 5.
    (3) Electronic systems.

(4)	(5)	(6)	(7)	(8)	(9)	(10)	(11)	(12)	(13)
Process	Progress	Number	per cent	Workers per group	Pilot	Size	Corp.	Tech- nology	Years
MS	IC	14	100	9	No		C	H	8

22. (1) *Fords (Finsbury) Ltd*
    (2) Chantry Avenue, Kempton, Bedford, HR2 7RS.
    (3) Aluminium closure making machines.

(4)	(5)	(6)	(7)	(8)	(9)	(10)	(11)	(12)	(13)
Process	Progress	Number	per cent	Workers per group	Pilot	Size	Corp.	Tech- nology	Years
M	PI	2	20		No			M	

23. (1) *V and E Friedland Ltd*
    (2) Reddish, Stockport, SK5 6BR, Cheshire.
    (3) Chimes, door bells, industrial alarm bells.

(4)	(5)	(6)	(7)	(8)	(9)	(10)	(11)	(12)	(13)
Process	Progress	Number	per cent	Workers per group	Pilot	Size	Corp.	Tech-nology	Years
M.A.	IC	12	100	6	No	80	1	M	

24. (1) *Garrick Engineering Ltd*
    (2)
    (3) Sub-contract machining.

(4)	(5)	(6)	(7)	(8)	(9)	(10)	(11)	(12)	(13)
Process	Progress	Number	per cent	Workers per group	Pilot	Size	Corp.	Tech-nology	Years
M	IC	4	100	6	No	24	1	M	1

25. (1) *GEC Elliott Ltd (Fisher Governor)*
    (2) Airport Works, Rochester, Kent.
    (3) Engine governors and automatic fluid control equipment

(4)	(5)	(6)	(7)	(8)	(9)	(10)	(11)	(12)	(13)
Process	Progress	Number	per cent	Workers per group	Pilot	Size	Corp.	Tech-nology	Years
M	PI	17			Yes		C	H	2

26. (1) *GEC Machines Ltd*
    (2) Thornbury, Bradford, Yorkshire.
    (3) Electric motors.

(4)	(5)	(6)	(7)	(8)	(9)	(10)	(11)	(12)	(13)
Process	Progress	Number	per cent	Workers per group	Pilot	Size	Corp.	Tech-nology	Years
M	PI	4	20	8	Yes	200	C	M	1

27. (1) *GKN Fasteners Ltd*
    (2) 26 Chase Road, London, NW10
    (3) Screws and fasteners.

(4)	(5)	(6)	(7)	(8)	(9)	(10)	(11)	(12)	(13)
Process	Progress	Number	per cent	Workers per group	Pilot	Size	Corp.	Tech-nology	Years
M	PI	4	4		Yes		C	M	

28. (1) *H. J. Godwins Ltd.*
    (2) Quenington, Cirencester, Gloucestershire.
    (3) Pumps, pistons, etc.

(4)	(5)	(6)	(7)	(8)	(9)	(10)	(11)	(12)	(13)
Process	Progress	Number	per cent	Workers per group	Pilot	Size	Corp.	Tech-nology	Years
M	IC	3	20	3	No	50	1	M	1

29. (1) *Hall, Thermotank International Ltd.*
    (2) 243, Vauxhall Bridge Road, London, SW1.
    (3) Refrigerating and air conditioning systems.

(4)	(5)	(6)	(7)	(8)	(9)	(10)	(11)	(12)	(13)
Process	Progress	Number	per cent	Workers per group	Pilot	Size	Corp.	Tech-nology	Years
M	PI							M	

30. (1) *T. S. Harrison Ltd*
    (2) Heckmondwike, Yorkshire.
    (3) Machine tools.

(4)	(5)	(6)	(7)	(8)	(8)	(10)	(11)	(12)	(13)
Process	Progress	Number	per cent	Workers per group	Pilot	Size	Corp.	Tech-nology	Years
M	PI	1	3	3	Yes		C	M	3

31. (1) *Hawker Siddeley Dynamics Ltd*
    (2) PO Box 34, Lostock, Bolton, Lancashire.
    (3) Aero engines.

(4)	(5)	(6)	(7)	(8)	(9)	(10)	(11)	(12)	(13)
Process	Progress	Number	per cent	Workers per group	Pilot	Size	Corp.	Technology	Years
A.M.	PI	3		15	No		C	H	5

32. (1) *Herbert Machine Tools Ltd*
    (2) Edgwick Works, PO Box 30, Coventry, Warwickshire.
    (3) Machine tools.

(4)	(5)	(6)	(7)	(8)	(9)	(10)	(11)	(12)	(13)
Process	Progress	Number	per cent	Workers per group	Pilot	Size	Corp.	Technology	Years
M.O.	IC	17	100	17	No	800	C	H	3

33. (1) *Hoover Ltd*
    (2) Western Avenue, Perivale, Greenford, Middlesex.
    (3) Vacuum cleaners and consumer durables.

(4)	(5)	(6)	(7)	(8)	(9)	(10)	(11)	(12)	(13)
Process	Progress	Number	per cent	Workers per group	Pilot	Size	Corp.	Technology	Years
M	PI						C	M	

34. (1) *Hopkinsons Ltd*
    (2) Britannia Works, Birkby, Huddersfield, Yorkshire.
    (3) Valves.

(4)	(5)	(6)	(7)	(8)	(9)	(10)	(11)	(12)	(13)
Process	Progress	Number	per cent	Workers per group	Pilot	Size	Corp.	Technology	Years
M	PI	1			Yes		I	M	

35. (1) *The Hydrovane Compressor Co Ltd*
    (2) Claybrook Drive, Washbrook Industrial Estate, Redditch.
    (3) Air compressors.

(4)	(5)	(6)	(7)	(8)	(9)	(10)	(11)	(12)	(13)
Process	Progress	Number	per cent	Workers per group	Pilot	Size	Corp.	Tech-nology	Years
M	PI	1			Yes		I	M	

36. (1) *Hyster (UK) Ltd*
    (2) Irvine, Ayrshire, Scotland.
    (3) Fork lift trucks.

(4)	(5)	(6)	(7)	(8)	(9)	(10)	(11)	(12)	(13)
Process	Progress	Number	per cent	Workers per group	Pilot	Size	Corp.	Tech-nology	Years
M	PI	18			No		C	M	3

37. (1) *Ingersoll Rand Ltd*
    (2) Queensway, Team Valley Trading Estate, Gateshead.
    (3) Compressors and pumps.

(4)	(5)	(6)	(7)	(8)	(9)	(10)	(11)	(12)	(13)
Process	Progress	Number	per cent	Workers per group	Pilot	Size	Corp.	Tech-nology	Years
M	PI	3	10		No		C	M	

38. (1) *International Engineering Ltd (ICL)*
    (2) Montgomery Road, Castlereagh, Belfast, Northern Ireland.
    (3) Computers.

(4)	(5)	(6)	(7)	(8)	(9)	(10)	(11)	(12)	(13)
Process	Progress	Number	per cent	Workers per group	Pilot	Size	Corp.	Tech-nology	Years
M	IC	9			No		C	H	

39. (1) *Jones and Shipman Ltd* (*Main Works*)
    (2) Narborough Road South, Leicester, LE3 2LF.
    (3) Grinding machines.

(4)	(5)	(6)	(7)	(8)	(9)	(10)	(11)	(12)	(13)
Process	Progress	Number	per cent	Workers per group	Pilot	Size	Corp.	Tech-nology	Years
M.A.	PI	3			No		C	H	4

40. (1) *Jones and Shipman Ltd* (*Small Tools*)
    (2) Narborough Road South, Leicester, LE3 2LF.
    (3) Small tools.

(4)	(5)	(6)	(7)	(8)	(9)	(10)	(11)	(12)	(13)
Process	Progress	Number	per cent	Workers per group	Pilot	Size	Corp.	Tech-nology	Years
M	IC	6			No		C	M	4

41. (1) *George Kent Ltd*
    (2) Biscot Road, Luton, Bedfordshire.
    (3) Precision measuring equipment.

(4)	(5)	(6)	(7)	(8)	(9)	(10)	(11)	(12)	(13)
Process	Progress	Number	per cent	Workers per group	Pilot	Size	Corp.	Tech-nology	Years
A	IC		100					H	

42. (1) *Linotype and Machinery Ltd*
    (2) Woodfield Road, Broadheath, Altrincham, Cheshire.
    (3) Typesetting machines.

(4)	(5)	(6)	(7)	(8)	(9)	(10)	(11)	(12)	(13)
Process	Progress	Number	per cent	Workers per group	Pilot	Size	Corp.	Tech-nology	Years
M	PI	6	10		No		C	H	

43. (1) *Mather and Platt Ltd.*
    (2) Park Works, Newton Heath, Manchester 10.
    (3) Electric motors, fire fighting equipment.

(4)	(5)	(6)	(7)	(8)	(9)	(10)	(11)	(12)	(13)
Process	Progress	Number	per cent	Workers per group	Pilot	Size	Corp.	Tech-nology	Years
M	IC	10		6	No		I	M	10

44. (1) *Matrix Machine Tools Ltd*
    (2) PO Box 39, Fletchampstead Highway, Coventry.
    (3) Machine tools.

(4)	(5)	(6)	(7)	(8)	(9)	(10)	(11)	(12)	(13)
Process	Progress	Number	per cent	Workers per group	Pilot	Size	Corp.	Tech-nology	Years
M	PI	1			Yes		C	H	1

45. (1) *Thomas Mercer Ltd*
    (2) Eywood Road, St. Albans, Hertfordshire.
    (3) Dial gauges.

(4)	(5)	(6)	(7)	(8)	(9)	(10)	(11)	(12)	(13)
Process	Progress	Number	per cent	Workers per group	Pilot	Size	Corp.	Tech-nology	Years
M	IC	14	50	6	No	202	I	H	7

46. (1) *Nelco Ltd*
    (2) Station Road, Shalford, Guildford, Surrey.
    (3) Electric motors, commutators.

(4)	(5)	(6)	(7)	(8)	(9)	(10)	(11)	(12)	(13)
Process	Progress	Number	per cent	Workers per group	Pilot	Size	Corp.	Tech-nology	Years
M	PI	3			No		I	M	

47. (1) *Newage Engineering Co.*
    (2) Aldbourne Road, Coventry, Worcestershire.
    (3) Gearboxes.

(4)	(5)	(6)	(7)	(8)	(9)	(10)	(11)	(12)	(13)
Process	Progress	Number	per cent	Workers per group	Pilot	Size	Corp.	Tech-nology	Years
M	IC	8	100		No			M	1

48. (1) *Platt Saco Ltd*
    (2) Bradley Fold, Bolton, Lancashire.
    (3) Gears and sheet metal parts.

(4)	(5)	(6)	(7)	(8)	(9)	(10)	(11)	(12)	(13)
Process	Progress	Number	per cent	Workers per group	Pilot	Size	Corp.	Tech-nology	Years
M.S.	PI	29	10	14	No	700	C	M	3

49. (1) *F Pratt Ltd*
    (2) Halifax, Yorkshire.
    (3) Chucks for machine tools.

(4)	(5)	(6)	(7)	(8)	(9)	(10)	(11)	(12)	(13)
Process	Progress	Number	per cent	Workers per group	Pilot	Size	Corp.	Tech-nology	Years
M	PI								

50. (1) *Quinton Hazel*
    (2) Colwyn Bay, North Wales.
    (3) Parts for motor vehicles.

(4)	(5)	(6)	(7)	(8)	(9)	(10)	(11)	(12)	(13)
Process	Progress	Number	per cent	Workers per group	Pilot	Size	Corp.	Tech-nology	Years
M	PI								

51. (1) *Rank Xerox Ltd*
    (2) Mitcheldean, Gloucestershire.
    (3) Copying machines.

(4)	(5)	(6)	(7)	(8)	(9)	(10)	(11)	(12)	(13)
Process	Progress	Number	per cent	Workers per group	Pilot	Size	Corp.	Tech-nology	Years
M	IC				No		C	M	

52. (1) *Rolls Royce* (1971) *Ltd*
    (2) Carrowreach Road, Dundonald, Belfast.
    (3) Precision parts for aeroengines.

(4)	(5)	(6)	(7)	(8)	(9)	(10)	(11)	(12)	(13)
Process	Progress	Number	per cent	Workers per group	Pilot	Size	Corp.	Tech-nology	Years
M	PI				No	255		H	

53. (1) *HM Royal Ordnance Factory*
    (2) Enfield Lock, Middlesex.
    (3) Small arms.

(4)	(5)	(6)	(7)	(8)	(9)	(10)	(11)	(12)	(13)
Process	Progress	Number	per cent	Workers per group	Pilot	Size	Corp.	Tech-nology	Years
M	PI				No		C	H	

54. (1) *Rubery Owen Ltd*
    (2) PO Box 36, Cleverleys Road, Warrington.
    (3)

(4)	(5)	(6)	(7)	(8)	(9)	(10)	(11)	(12)	(13)
Process	Progress	Number	per cent	Workers per group	Pilot	Size	Corp.	Tech-nology	Years
M	PI				Yes				

55. (1) *Serck Audco Valves Ltd*
    (2) Newport, Shropshire.
    (3) Valves.

(4)	(5)	(6)	(7)	(8)	(9)	(10)	(11)	(12)	(13)
Process	Progress	Number	per cent	Workers per group	Pilot	Size	Corp.	Technology	Years
M	IC	24	100	18	No	1000	C	M	14

56. (1) *Sigmund Pulsometer Ltd*
    (2) Oxford Road, Reading, Berkshire.
    (3) Pumps.

(4)	(5)	(6)	(7)	(8)	(9)	(10)	(11)	(12)	(13)
Process	Progress	Number	per cent	Workers per group	Pilot	Size	Corp.	Technology	Years
M	PI	2			Yes		I	M	8

57. (1) *Simon Container Machinery Ltd*
    (2) Birdhall Lane, Cheadle Heath, Stockport, Cheshire.
    (3) Printing, container and flour milling machines.

(4)	(5)	(6)	(7)	(8)	(9)	(10)	(11)	(12)	(13)
Process	Progress	Number	per cent	Workers per group	Pilot	Size	Corp.	Technology	Years
M	IC	9		25	No	430	C	M	4

58. (1) *Simon – VK Ltd* (*Victory–Kidder*)
    (2) Arrowbrook, Upton, Wirral, Cheshire.
    (3) Printing machines.

(4)	(5)	(6)	(7)	(8)	(9)	(10)	(11)	(12)	(13)
Process	Progress	Number	per cent	Workers per group	Pilot	Size	Corp.	Technology	Years
M	IC	5			No		C	M	

59. (1) *Smiths Industries Ltd* (*Aviation Division*)
    (2) Bishops Cleve, Cheltenham, Gloucestershire.
    (3) Instruments.

(4)	(5)	(6)	(7)	(8)	(9)	(10)	(11)	(12)	(13)
Process	Progress	Number	per cent	Workers per group	Pilot	Size	Corp.	Technology	Years
M	PI	2	5	7	No		C	H	3

60. (1) *Timson Ltd*
    (2) Perfecta Works, Kettering, Northamptonshire.
    (3) Printing machinery.

(4)	(5)	(6)	(7)	(8)	(9)	(10)	(11)	(12)	(13)
Process	Progress	Number	per cent	Workers per group	Pilot	Size	Corp.	Technology	Years
M	PI				No			M	

61. (1) *Triangle Valve Co.*
    (2)
    (3) Valves.

(4)	(5)	(6)	(7)	(8)	(9)	(10)	(11)	(12)	(13)
Process	Progress	Number	per cent	Workers per group	Pilot	Size	Corp.	Technology	Years
M	PI								

62. (1) *Vickers Ltd* (*Engineering Group*)
    (2) South Marston Works, Swindon, Wiltshire.
    (3) Valves.

(4)	(5)	(6)	(7)	(8)	(9)	(10)	(11)	(12)	(13)
Process	Progress	Number	per cent	Workers per group	Pilot	Size	Corp.	Technology	Years
M	PI	1	2		Yes	425	C	M	

63. (1) *Walker Crossweller and Co.*
    (2) Whaddon Works, Cromwell Road, Cheltenham, Gloucestershire.
    (3) Valves for hot water mixing.

(4)	(5)	(6)	(7)	(8)	(9)	(10)	(11)	(12)	(13)
Process	Progress	Number	per cent	Workers per group	Pilot	Size	Corp.	Tech-nology	Years
M	PI	3		6	No	271	I	M	4

64. (1) *H. W. Ward and Co. Ltd*
    (2) Bilford Road, Blackpole, Worcestershire.
    (3) Machine Tools.

(4)	(5)	(6)	(7)	(8)	(9)	(10)	(11)	(12)	(13)
Process	Progress	Number	per cent	Workers per group	Pilot	Size	Corp.	Tech-nology	Years
A	PI	2		7	No	403		H	

65. (1) *G and J Weir Ltd*
    (2) Carthcart, Glasgow.
    (3) Pumps.

(4)	(5)	(6)	(7)	(8)	(9)	(10)	(11)	(12)	(13)
Process	Progress	Number	per cent	Workers per group	Pilot	Size	Corp.	Tech-nology	Years
M	PI				Yes			M	

66. (1) *Westinghouse Brake and Signal Co. Ltd*
    (2) Chippenham, Wiltshire.
    (3) Railway Equipment.

(4)	(5)	(6)	(7)	(8)	(9)	(10)	(11)	(12)	(13)
Process	Progress	Number	per cent	Workers per group	Pilot	Size	Corp.	Tech-nology	Years
M	IC	7			No	190	C	M	3

67. (1) *Whittaker Hall (Fluidair compressors) Ltd*
    (2) Miller Street, Radcliffe, Manchester.
    (3) Air compressors.

(4)	(5)	(6)	(7)	(8)	(9)	(10)	(11)	(12)	(13)
Process	Progress	Number	per cent	Workers per group	Pilot	Size	Corp.	Tech-nology	Years
M.A.	IC	6	100	5	No	50	I	M	7

INDEX